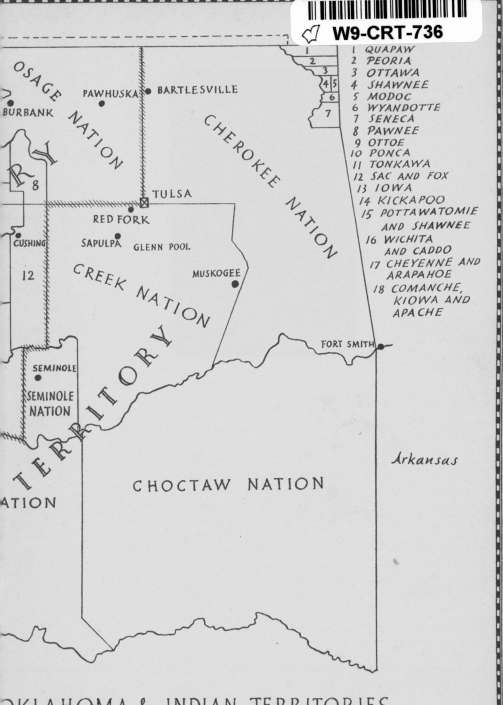

1 QUAPAW
2 PEORIA
3 OTTAWA
4 SHAWNEE
5 MODOC
6 WYANDOTTE
7 SENECA
8 PAWNEE
9 OTTOE
10 PONCA
11 TONKAWA
12 SAC AND FOX
13 IOWA
14 KICKAPOO
15 POTTAWATOMIE
   AND SHAWNEE
16 WICHITA
   AND CADDO
17 CHEYENNE AND
   ARAPAHOE
18 COMANCHE,
   KIOWA AND
   APACHE

OSAGE NATION

PAWHUSKA   BARTLESVILLE

BURBANK

CHEROKEE NATION

TERRITORY

8

TULSA

RED FORK

SAPULPA   GLENN POOL

CUSHING

12

CREEK NATION

MUSKOGEE

TERRITORY

SEMINOLE

SEMINOLE NATION

FORT SMITH

Arkansas

CHOCTAW NATION

NATION

OKLAHOMA & INDIAN TERRITORIES
1890

THEN CAME OIL

By C. B. GLASSCOCK

The "Wild Mary" Sudik, probably the most widely publicized oil gusher in the world. Black spray from this well drenched Oklahoma City and miles of countryside.

THEN CAME OIL

OIL

The Story of the Last Frontier

———★———

By C. B. GLASSCOCK

THE BOBBS-MERRILL COMPANY

Indianapolis   *Publishers*   New York

FIRST EDITION

Printed in the United States of America

PRINTED AND BOUND BY
BRAUNWORTH & CO., INC.
BUILDERS OF BOOKS
BRIDGEPORT, CONN.

CONTENTS

ILLUSTRATIONS

THEN CAME OIL

THEN CAME OIL

CHAPTER I

THE TRAIL OF TEARS

FOR three centuries the white man, in the name of God, in the name of Civilization, in pursuit of liberty, fought his way through the wilderness of America. In blood and sweat the frontiers fell before him until only one remained. Chance and circumstance and governmental policy had left one great area, bounded by Kansas, Arkansas and Texas, comparatively untouched.

Then came oil. In a tenth of the time that the United States had been aware of the frontiers as limitless lands of opportunity for hardy souls, the last frontier was destroyed. Monumental buildings arose upon the sites of teepees. Universities spread their influence over the ranges of the buffalo. Cocktails were dispensed from chromium-trimmed bars in luxurious homes. The calumet gave way to the cigarette. Civilization rather than the cowboy was in the saddle, upon the last wide, free and dangerous land within the nation.

It was a large and important area, the size of a great state, the state of Oklahoma. It was an impressive part of the original frontier which had existed, diminishing through three hundred years, as a challenge and a reward. The conquest of that greater territory had built the nation, given strength and pride to its people, romance and color to its history. The formation and conquest of the last frontier, and the changes which the coming of oil wrought upon it, is a drama in our national life.

A prelude to that drama began with what has been chronicled in the records and legends of the once-great tribe of the Cherokees as "The Trail of Tears." That was a trail from the village of New Echota, Georgia, capital of the self-contained and civilized Cherokee Nation a century ago, to a land of exile hundreds of miles away in the western wilderness near the Arkansas River. It was a trail to be marked by the graves of hundreds of their people, graves more closely spaced through a thousand winding miles than the highway route numbers are spaced today.

Fate had long conspired against the Cherokees as it had against their neighboring nations, Creek, Choctaw, Chickasaw and Seminole, in Georgia, Alabama, Tennessee, Mississippi and Florida. All those tribes were herded on journeys of despair and death from their homelands in half a dozen states to a wild and worthless land three hundred miles and more west of the Mississippi. The tragedy of the Cherokees may be cited as revealing the tragedy of all.

After the American Revolution the commonwealth of Georgia laid claim to all lands between the parallels of its northern and southern borders extending westward to the limits of United States jurisdiction. Most of those lands were uninhabited except by Indians. To give theoretical control of those unexplored regions to the Federal government, Georgia agreed to a quitclaim of the area beyond its present western border, with the understanding that the Indians within the state should be removed.

That understanding was reached in 1802, without consulting the Indians. Most of the Indians had sided with the English against the colonies. It was the sense of Georgia and of numerous other states that evacuation of their lands was a suitable punishment. Few of the Indians were even aware of the arrangement. Few protested, and they had little influence.

One chastened group of Cherokees returning from Washington toward their homes found themselves tricked and robbed by a company of white traders on the Tennessee River. And promptly they slaughtered the whites. Though partly civilized by long contact with the English colonists, the Cherokees had not yet reached the status to which their nation was to come. Their homes and lands in Georgia were not yet developed into rich plantations. Only a few owned slaves. Some had broken their promise to be good Indians, a nation friendly to the United States, almost as soon as it was made. They feared punishment. Instead of returning to their homes they turned westward into Arkansas and established themselves, becoming known as the Western Cherokees. Some of their Georgian and Carolinian people joined them.

But the larger part of the nation remained within northern Georgia. They planted, built, and prospered. They bought slaves and improved plantations. The famous Sequoyah invented an Indian alphabet and had it cast in type. They printed a newspaper in their native language. They maintained their law, their courts and their schools as a separate nation, recognized to the extent of treaties with the government of the United States. Many were educated in English. Their homes were furnished in comfort. Music and books were available. Black slaves waited upon them in the same manner and with the same loyalty with which the blacks served the whites.

That was the nation in which Tsan Ross or John Ross arose to the office of principal chief, planning to make his nation a state within the Union. Ross was a brilliant man, highly educated. He was aware through blood and training of all the finer traditions of his people, and aware through education of all the best that the white civilization had to offer. Andrew Jackson, President of the United States, was his familiar friend. Henry Clay, Daniel Webster, Sam Houston, Davy Crockett, and others of the leading men in American history were his friends and admirers.

For ten years as chief of the Cherokees he fought their battles against the encroachment of the commonwealth of Georgia, through all the fields available under the Federal constitution and law. He forced from President Jackson in the presence of some of the nation's most noteworthy men the admission that it was the duty of the Federal government to protect the rights of the Cherokees at any cost. He won a decision from the Supreme Court of the United States against a decision of the Georgia legislature expelling his people from their homes and plantations within Georgia. And then he saw the President sacrifice principle to political expediency with the statement that Marshall had rendered the decision, let him enforce it.

Through a third of a century of such struggles the Cherokees had grown prosperous, more civilized, more attached to their homes, more united as a nation than they had ever been in the days of their savagery. A new generation had arisen since the commonwealth of Georgia and the government of the United States had agreed to drive them from their lands. The new generation had inherited and improved those lands. The possibility of enforced removal was as far beyond their conception as an order for the removal of all the people of Georgia to the wilds of Siberia would be today.

Then came gold! Georgians were chafing against the stubborn defiance of the Cherokees when gold was discovered near the Cherokee capital. The same lust for gold which had brought DeSoto and his *caballeros* into the Southwest three centuries earlier, which had started the conquest of America, came again to complete the ruin of the Indians.

Even the wise and witty Davy Crockett, scout, soldier, businessman, Congressman from Tennessee, loyal friend of the Cherokees, was trapped momentarily in the rush for gold upon the Indian lands. Digging foundations for a mill by agreement with the Cherokees to extend the business which had made him prosperous in his own state, Crockett was seized by Georgia soldiers, handcuffed and thrown into jail. There he learned of the restrictions arbitrarily imposed upon the Cherokees within their own nation by the commonwealth of Georgia. No man might dig, even his own ground, without special permission from the state.

The fact that Crockett was a congressman from a sister state, digging foundations for a mill, was no extenuation. No man could dig for anything on Cherokee land. Georgia had decreed it. Grinning sardonically, Davy Crockett demanded to know how the Indians might get their potatoes out of the ground, how they might bury their dead. He managed to talk his way out of jail, but otherwise obtained no satisfaction. Sensing the impending tragedy, he warned the Indians and carried their battle to Washington, to President Jackson, the friend under whom he had fought valiantly in western Indian wars.

His efforts were futile, as were the appeals of John Ross, Sam Houston and others of the great men of the day. President Jackson preferred to placate the Georgians. States' rights were a living force in those days. If the Federal government did not evacuate the Cherokees, Georgia would do so. Let justice and the Indians fall where they might.

A land-lottery system was instituted to confiscate the plantations and homes of the Cherokees. Leading men of the tribe returned to their homes to find them occupied by new "owners" whom they had never seen but who were protected in their occupancy by guns in the hands of the militia. Poorer members of the tribe found their garden patches, their chief source of livelihood, confiscated and destroyed by gold-seekers under the protective guns of armed state guards.

John Ross himself was one of the first men driven with

his family from his luxurious home, to a cabin across the Tennessee line to await and to organize the tragic exodus into a wild and distant land.

Federal soldiers under General Winfield Scott, assigned by President Jackson to complete the roundup and move the despairing nation on its way, joined with the Georgia militia in the task. Some fifteen thousand Cherokees were torn from their homes. A few hundred managed to escape the soldiery, and found refuge in the neighboring hills, preferring to suffer any hardship in the country of their ancestors rather than move to distant unknown lands with their despairing people. Descendants of those Cherokees live in that region still. Groups of hundreds, of a thousand and more, were sent forth on various routes, partly by water and partly by land, losing ten per cent of their number by death on the way. The greater number were herded into stockades to await the mass movement.

The year was 1838. There were what passed as roads through parts of Tennessee, Missouri and Arkansas. There were towns and farms. But those very adjuncts of civilization made the journey more terrible. They had reduced the natural game supply. The thousands of men, women and children in the throng could not take a living from the country as they moved. They could not carry provisions for a winding journey of near a thousand miles. There was no god of the Indians as there was a God of Israel to guide and guard and feed the wanderers.

It was autumn before the Cherokees were moved from the stockades in which they had been imprisoned to a central encampment in eastern Tennessee. Even the prospect of a journey on foot through a thousand miles of unknown country, through the floods and snows of winter, appeared better by that time than their recent sufferings, ill-housed, ill-fed, without sanitation, despairing in idleness, with hundreds dying, in the stockades.

As quickly as might be the caravan was organized by

General Scott. The train was divided into sections, miles long, formed with as much military precision as the able veteran of the War of 1812 and of half a dozen Indian wars could effect. Camp equipment, food, and meager personal belongings were carried in wagons, each of which was the center of a small unit of the great caravan. Each unit so formed was charged with the responsibility of keeping its equipment in repair, and maintaining its place. The main train of wagons, carrying reserve supplies for the entire caravan and the troops, was placed near the center of the line. Mounted soldiers were assigned to maintain pressure upon the lagging rear and upon the flanks, to round up stragglers and runaways.

There were hundreds of sick, and aged, and children hardly able to walk. There were men and women bowed down with burdens which the insufficient wagons could not carry. There were slaves, household and plantation retainers, mostly the old and feeble or the young and help-less who had not been confiscated or assigned. They forced their way into the line, seeking their masters, pleading to be taken to the new home of the nation in the West.

Days were consumed in the preliminary task of break-ing up the central encampment and moving the Indians out to their specified places in the line. Whips cracked. Goads prodded. Horses threw themselves into their collars. Oxen settled against their yokes. The long journey began.

Never in the history of America, perhaps never in the history of the world, has there been a scene to compare with it. On this continent the expulsion of the Acadians from Nova Scotia as immortalized in Longfellow's *Evangeline* presents a slightly similar tragedy in miniature. Possibly the exodus of the Israelites from Egypt, of all historic or legendary tales of united movement of a race from land to land, is best comparable. But the Israelites moved of their own volition toward a Promised Land flowing with milk and honey, sanctified by the word of Jehovah, the land of their ancestors, of Abraham, of Isaac and of Jacob. Theirs was a journey out of bondage, a journey of hope, sustained by their God.

The Cherokees moved not from bondage but from freedom. They also moved toward a promised land, but one promised by the white men from whom they had learned to expect only lies. They had no illusions that the descriptions of a haven of peace and plenty made to them by the United States government would prove to be true. Their own chief, John Ross, had visited that land and reported it as a region of desolation. Their trusted friend, Sam Houston, rising before President Jackson in the final appeal for justice, had said that the western land was no such haven of safety and prosperity as had been pictured by the government. He had visited it. He warned that the civilized Cherokees would find themselves surrounded by savagery with which they could not cope. That was the promised land toward which the people plodded, starving, storm-worn, desperate.

A bitter winter closed in upon them. Slowly they moved through Tennessee. Ten miles was a long day's journey through mud and snow when fifteen thousand persons were traveling afoot, requiring time to bury their dead at every stop, to care for their dying, to break a far-flung encampment with the population of a city, to make a new camp before the coming of darkness. Rivers and storms and sickness and death halted them completely at times. All the bitter months of that long winter and the dusty heat of a following summer were not enough to complete the journey.

But at last the broken remnants of the tribe arrived to establish their nation anew upon the northern side of the Arkansas river, with a broad outlet westward to the buffalo ranges. In approximately the same period, and over ways of similar hardship, came the other civilized tribes. To the Creeks and Seminoles had been assigned a similar area north of the Canadian River, extending westward to the Texas Panhandle. The Choctaw and Chickasaw nations occupied the area between the Canadian and Red rivers from Arkansas to Texas, roughly a third of the present state of Oklahoma.

The Five Civilized Tribes had been moved over their trails of tears into a wilderness. Savage tribes still roamed it, but by treaty with a white authority which possessed the power, if not the right, it was assigned to have and to hold by the five tribal nations, unmolested by the whites, "as long as water flows and grass grows."

In their simplicity they hoped. Almost, they believed. But another century stretched before them.

CHAPTER II

The Frontier Closes In

The American frontier was moving westward with far greater speed and force in the early years of the nineteenth century than it had been moving through the two hundred years after the first settlements on the Atlantic coast. The movement strengthened and accelerated as floods gain power with the weight of waters behind them and within them.

The Six Nations of the Iroquois, and others of the savage eastern tribes, had been almost destroyed by the manifest destiny of the whites. They had taken heavy toll in blood and terror, but it had not been enough. The whites moved on, across the Alleghenies, across the Ohio River, through the Great Lakes basin, into Ohio, Michigan, Indiana, Illinois, into Kentucky, Tennessee, Alabama—ever westward.

Most of those pioneers were a rude and violent people, almost as rude and violent as the savage denizens of the regions they sought to occupy. And they had the great practical advantage of a background of centuries of civilization. They sought homes and farms and freedom. Their advance guards of explorers and trappers and traders revealed the way, the dangers and hardships and rewards.

They accepted the dangers and hardships, and pushed on for the rewards. They traded violence for violence, treachery for treachery, scalps for scalps, with the Indians. Always, more or less aided in various ways by their government, they pushed onward, driving the tribes before them.

20

France in 1803 had ceded to the United States for $15,000,000 the vast area known as Louisiana, west of the Mississippi, north of the Gulf, including all the great valley of the Missouri and its tributaries. The United States knew it only as a wild land occupied by savage tribes but with immediate potential riches in the form of furs available for such hardy trappers and traders as dared defy its dangers.

But with its acquisition, and the steady pressure of land-hungry pioneers from the eastward, the practice of two centuries of white civilization against the red men began to take form as a policy. There was a wild crude land into which the Indians who had long been raiding, burning and murdering among the conquerors of their homelands might be driven and restricted to work out their own destiny.

All the country east of the Mississippi, which then included all the states and territories of the Union, could thus be made safe for the white man. The last remnants of the Iroquois and other eastern tribes in Ohio, Michigan and Indiana were moved by so-called treaty, or by military force, across the Mississippi and the Missouri. All the vast country lying west of the Arkansas and Missouri boundaries and the Missouri River, "as far west as the country is inhabitable," was designated as Indian country.

The fact that great sections of the richest part of the area were already occupied by savage tribes, and larger sections were common hunting grounds of various tribes, seemed a minor point. The various native tribes were no more friendly or trustworthy in their relations with each other than they were with the encroaching whites. With the exception of a few noteworthy leaders stupid savagery was their outstanding characteristic. Deception, when it could be turned to the accomplishment of any purpose, was its own justification, traditionally a sign of shrewdness, never of dishonor. There were no words in the languages of the western Indians for what we know as morality in its broader sense.

Treachery was practiced by the whites but it was not introduced by the whites. It was well known and widely practiced by the Indians long before they ever saw a white man. Savage warfare was their most honored occupation. Next to that came the theft of horses as an activity in which a young savage could win wealth and glory for himself and his tribe. Comanches and Pawnees and Osages among the leading western tribes were hereditary enemies. No ritualistic dance of mourning at an Osage funeral was complete without the taking of the scalp of an enemy. Choctaws on their western hunting trips had battled fiercely with the Caddoes. Western Cherokees, less softened and civilized than the eastern portion of their tribe, warred with the Osages. The Quapaws were as savage.

Yet all the Indians coveted the goods brought by white traders, the knives, guns, beads, blankets, gaily colored calicos, whiskey. For those they would trade their furs and skins, even their lands.

Treaty-making was a time-honored practice among the Indians. They delighted in its formalities and in the feasting and dancing between tribal raids and wars. For countless generations they had been digging up and wielding the hatchet chiefly that they might enjoy the celebration of its reburial. They entered willingly into negotiations with the whites.

So, beginning with one treaty in 1818 and concluding with another in 1825, the Osages ceded to the Federal government their claims to lands in the present Oklahoma in return for cash payments and other items of value, and retired to a more restricted region in southern Kansas. The Quapaws by treaties in 1818 and 1830 ceded a broad area along the Red River. Comanches and other tribes native to the plains made similar agreements.

Demoralization of the western Indians with whiskey, gifts, and free rations with consequent idleness was well under way. When Thomas James and John McKnight established their first temporary trading post a few miles west of what is now Oklahoma City the Comanches promptly robbed and murdered McKnight. With the naïveté of children they then sent word to James that his post was too close to the Osages for them to venture there to trade. Only the year before a war party of those same Comanches had held up the first James expedition and exacted a tribute of nearly all the trade goods in return for a promise of safe passage. Only a few weeks later, with the scalp of McKnight hardly dry at a savage's belt, the Comanches invited James to their village to trade.

Similar incidents were being enacted throughout the Indian territory. Treaties with the Osage and Kansas Indians authorized by Congress in 1825 for safe passage of what was to be defined as the Santa Fe Trail from the Missouri frontier marked the beginning of the end. The trail was roughly marked, and for a time trading caravans traveled it with a reasonable degree of safety. Settlers soon followed the traders into the rich Kansas prairies. Military posts were pushed westward for their protection. Fort Smith, Fort Scott, Fort Reilly, Fort Gibson and a dozen others were established at the border and within the Indian country.

Each of the so-called civilized tribes established its tribal government, tilled its corn fields and squash patches, herded its cattle, maintained its villages, hunted upon the buffalo ranges and gave battle to its savage neighbors. That was the situation through approximately the second quarter of the century.

Incidentally it might be noted in that connection that the negro slaves of the Five Civilized Tribes did most of the manual labor of the agricultural development. As early as 1836 Albert Gallatin stated that the number of plows in use among the Five Tribes answered for the number of able-bodied negroes. As the years went on, up to Lincoln's emancipation of the slaves, similar conditions obtained. Indian men had never taken kindly to manual labor. It cannot be doubted that much of the economic and cultural superiority of the Five Tribes was due to their adoption of the institution of negro slavery. It enabled them to put the improved agricultural methods of the whites into practice. It made them less dependent than the savage tribes upon hunting and raiding for a living. It made them less nomadic, and resulted in improvement in the quality of their homes.

A brief new era began in 1854. President Pierce signed the Kansas-Nebraska Bill, making Kansas and Nebraska each a territory of the United States. The Indian territory was correspondingly reduced to a fraction of its original vast area. It was limited approximately to the area of the Oklahoma of today with the exception of its Panhandle. The Federal policy of restricting Indians within an area where they might be controlled, by force if necessary, leaving the more inviting lands to the whites, had been definitely established.

The migration of the Mormons from Illinois to the Great Salt Lake basin had indirectly revealed new possibilities of wealth, agriculture and prosperity to eastern America. Texas had joined the Union. The vast Oregon country of the Northwest had been taken over from Great Britain. The gold rush to California had further impressed the East with new values in the Far West. A new wave of pioneers swept onward, leaped the Rockies, the desert and the Sierras, and began its backwash from the Pacific.

A war beside which the Seminole war in Florida and Georgia, the Creek war in Alabama or any of a score of major Indian wars were as child's play came upon the nation.

The Choctaw and Chickasaw tribes declared themselves as a unit for the Confederacy. Theirs were the traditions of the slave owners. They were residents of southern territory. The whites who had intermarried among them and lived as members of the tribes were of southern antecedents. Most of the Federal Indian agents who exerted considerable influence among them were for the South by conviction. And when the North abandoned its forts in the territory, leaving the populace to the mercies of the Confederacy, self-protection became a determining influence. Within a few months after the start of the war the Confederacy practically controlled the entire Indian territory.

The Cherokees had divided when two of their leaders, Stand Watie and E. C. Boudinot, accepted commissions in the Confederate army and carried numerous warriors with them, while the John Ross faction had moved to the North. Ross had never forgiven Georgia for the expulsion of his people. The Creeks divided for similar reasons. Two sons of the Chief McIntosh who had been executed for agreeing to removal accepted Confederate commissions and influenced their followers to join the South. Their traditional opponents, under the leadership of Chief Yo-ho-la, who had executed their father for selling out the tribe thirty years earlier, moved into Kansas. A large part of the Seminole tribe did likewise.

The death list among those refugees in the winter of 1861-1862 was appalling. Most of them had abandoned comfortable homes, cattle and even clothing in their flight northward. The Federal government was too busy with the war to bother about them. Exposure, sickness and starvation took a greater toll from the tribesmen loyal to the Union than did the guns of their brothers in the South.

Much of the territory held by the Five Tribes, which then constituted most of the area which is now Oklahoma, was devastated by the war. Notable progress in agriculture, with attendant advantages in civilization which had been accomplished within a generation, was virtually destroyed. And, as had been the result in scores of minor wars, the Indians again were defeated.

It served as an excuse if not a reason for renewed pressure upon them. Their numbers had been decimated. They needed, or the government argued that they needed, far less than the lands which had been accorded them a generation earlier. New treaties, the more important of which were negotiated in 1866, ceded most of the western Oklahoma lands back to the Federal government.

The Creeks surrendered the entire western half of their dominion, at a valuation of thirty cents an acre, with payment to be made when surveys had been accomplished. As in other treaties it was stipulated that these lands could be divided by the government into reservations for other Indian tribes. As a matter of fact a considerable part of the area, which was to become known as Old Oklahoma, was never assigned to any tribe but remained theoretically vacant until it was opened to settlement in the first great land rush of 1889.

Possibly because Union sentiment was stronger among the Cherokees, the post-war treaty stipulated that the government could establish only so-called civilized tribes within the Cherokee lands proper, on agreement with the tribe. Wild tribes could be settled in the Cherokee Outlet to the west by paying the Cherokees for such lands at a price to be determined later. Thus the Cherokees did not lose all their western lands at the moment. A vast acreage was purchased later, immediately prior to the opening of the Cherokee Strip to settlement in 1893. And thereby hangs another tale which will be told in its place.

That great area, 58 miles wide by 225 miles long, adjoined the southern border of Kansas. Discovery of a discrepancy in earlier official surveys and records disclosed an area 2.46 miles wide and 276 miles long, claimed alike by

Kansas and the Cherokees whose Outlet it bordered. This narrow strip, containing about 450,000 acres, was the Cherokee Strip proper which has so often been confused with the Cherokee Outlet and finally gave its name to the Outlet. The Outlet was originally assigned to the Cherokees as a way to the buffalo-hunting ranges.

Under an act of Congress prior to the treaties of 1866, arrangements had been made to include the narrow "Strip" in the state of Kansas. The claims of the Cherokees were to be satisfied by selling the land to settlers and investing the proceeds, amounting to $560,361, in government bonds, with income to be used for the Cherokees as needed.

The government's agreement with the Creeks, Seminoles and other tribes providing assignment of their ceded lands to other tribes was soon put into effect. Beginning promptly after the treaties of 1866, and continuing through another decade, various tribes from all sections of the vast original Indian territory extending from the northeast Texas border to the Canadian line were herded into the new reservations.

It was no easy matter. The plains Indians and the mountain Indians of the Dakotas, Montana and Idaho were fighting red devils. The Cheyennes and Sioux especially resisted with fanatic savagery. But they could not defeat manifest destiny, though they could claim ten thousand white men's scalps and the destruction of some thousands of farm homes and trading posts in the process.

Remnants of the Senecas who had made life miserable and hazardous for the colonists in the Mohawk Valley of New York a century earlier, and who had been pushed westward through the years, were settled at the northeast corner of the restricted Cherokee Nation. Delawares whose ancestors had sold Manhattan Island to the Dutch, and the site of Philadelphia to William Penn, and who had drifted through ten states in three centuries of white settlement, were established. Shawnees, Pawnees, Iowas, Kiowas, Comanches, Arapahoes, Cheyennes, Sac and Fox, and numerous others were assigned to reservations and generally held within bounds.

Largest of the tribes within the limits of Kansas at the close of the war was the Osage. For a century the very name of Osage was a terror to various tribes all the way from Michigan to the Red River. But beginning with their concession of Missouri and Arkansas territory to the United States in 1808, both the strength and terror of the tribe slowly faded. Federal provision for the tribe in payment for their cession of other lands, including Oklahoma, had sadly demoralized a nation of hunters and warriors.

It was not difficult to place them upon a comparatively tiny fraction of their original unlimited range, with a government pledge of protection and rations. Under that arrangement they could be maintained upon the most barren, desolate and apparently valueless lands within the territory. And such lands, between the Cherokee reservation and the Arkansas River, were assigned to them. Less than half a century was to prove that arrangement one of the most ironic practical jokes ever perpetrated by Fate.

Other tribes, notably the Cheyennes, were less amenable to control. When the spoils system of government replaced reliable and experienced Indian agents with political appointees who knew or cared nothing about the red men, trouble quickly developed. Cattle and horse stealing, murders, and punitive action by the sufferers became general. The Indian question became a pressing problem before a government already staggering under the huge debt of the War Between the States and muddling through the reconstruction of the devastated South. The government had virtually no money with which to feed and pacify the less civilized of the plains Indians, even if it had wanted to do so.

General Winfield Scott Hancock, veteran of the Mexican and Civil wars, undertook to settle the difficulties by force. He knew practically nothing of the Indian character or the background of the situation. In the spring of 1867 he led his troops into the Arkansas Valley in Kansas where the Cheyennes had incurred the enmity of the whites. When the Indians, traveling through snow on their winter-starved ponies, failed to assemble at his order as quickly as he believed they should, he berated them with the insolence of a military martinet. Then he proceeded to accuse them of innumerable misdeeds in which they had had no part.

Having thus aroused their bitter resentment, he announced that he would visit their village for inspection and further council. When the squaws and children vanished from the village before his approach he charged treachery. When they failed to return at his command he burned the village. Word of the injustice spread quickly among neighboring tribes. Tribesmen burned and scalped among the settlers. The warfare spread like a prairie fire. All through the summer and autumn the musket, the tomahawk, the bow, the scalping-knife and the torch took toll. With the approach of winter the Indians were gathered once more in peace council with the whites. Seven thousand of them sullenly assembled at Medicine Lodge River, near the southern Kansas border.

New treaties, new promises, were signed and sealed, for the moment, with Federal supplies to the starving natives. The last of the Kansas tribes were to settle in reservations of western Oklahoma. Flatly they were informed that the buffalo which had meant life to their ancestors for generations must give way to cattle. The Indians must settle down to farming, for which implements, seed and instruction would be furnished by the government. But this time there were to be no slaves to do the work. And in the following spring Congress, more interested in the reconstruction of the South, failed to make appropriation for Indian supplies. Again the savage warfare swept the West, led by the Cheyennes, who had been told, in effect, that they must look out for themselves.

With the coming of winter the Cheyennes returned to their villages along the Washita River. Other Indians also returned to their reservations. But the United States had had enough. It might deal again with the others, but never again with the Cheyennes. General George A. Custer, cavalry hero of the Civil War, was assigned the task of final punishment of the Cheyennes.

With a regiment of cavalry Custer moved out of Camp Supply late in November, 1868. His military mission was to punish the Cheyennes, and incidentally the Arapahoes, Kiowas and Comanches, for their depredations. The white men ignored the fact that the Indians had been moved to those depredations in part by starvation due to the government's failure to furnish rations according to the treaty pledges. The Indians were desperate as well as savage. The well-fed, well-armed, well-mounted troopers had every advantage. They moved as swiftly as the deep snow and bitter weather of an early winter would permit them toward the Washita River. There the Indians were known to have established winter camps.

It had never been the practice of the Indians to carry their warfare into the winter. Feed for their ponies and for themselves was too scarce for winter campaigns. Apparently they failed to understand that white warfare could be conducted on a different basis. The Custer troops were able to make a surprise attack upon the first village encountered. The event has gone down in history as the Battle of the Washita, and has even been cited as a notable victory.

In fact it was a massacre. Not a force of competent Indian warriors, but a winter-stricken village containing more women and children than fighting men, was the ob-

A modern oil pipe line offers striking contrast to the first transportation of oil, in barrels, in Oklahoma.

ject of the attack. Men, women and children alike were killed. Black Kettle, chief of the village, and his wife were shot down side by side. Almost every Indian who escaped the opening volleys fled to the stream and sought refuge under the banks, wading breast deep in the icy waters to escape the bullets of the troopers.

Many women and children were shot down as they attempted to cross the freezing stream or dodged in terror through the brush. Warriors fell while attempting to cover the retreat of their families. The village itself was burned to the ground by Custer's orders.

Custer's subsequent report to his superior, General Sheridan, claimed one hundred and three warriors slain, and fifty-three women and children captured. It failed to note what was probably a greater number of non-combatants slain. Major Elliott and a force of fifteen men detached by Custer to disperse a group of refugees emerging from the river at a point a mile or so below the village were the chief casualties of the white command. Surrounded by the fleeing fighting warriors, who included Arapahoes, Kiowas and Apaches from down-river villages as well as the Cheyennes, this detachment was killed to the last man.

Still it was "a famous victory." Custer prudently refrained from pressing it among the other villages whose warriors were pouring forth in numbers equal to the white troopers.

Friends of the Indians if not admirers of Custer may find in that massacre upon the Washita some poetic justification of the more famous Custer Massacre which was to end the life of the brave, vain and brilliant General only a few years later in the Battle of Little Big Horn. That battle was the beginning of the end of the last great uprising of the American Indians against the white conqueror.

That insurrection, led by Sitting Bull and Crazy Horse

of the Sioux, and supported by most of the lesser tribes of Dakota, Montana and Wyoming, had brought death and destruction to innumerable traders and settlers in the North and West. Beginning in 1873, Custer cleared the Black Hills country of the Indian menace. But Sitting Bull and Crazy Horse were leaders, warriors and tacticians as great or greater in their way than the soldiers of the United States who opposed them. They were to mark the end of the final chapter of organized Indian warfare in America.

CHAPTER III

THE FINAL INDIAN WARS

WITH the fierce Sioux making a final concerted stand against white encroachment in the wilder areas of the Northwest, the American frontier as a scene of warfare and brutality was briefly concentrated in that area. The major remnants of a score of Indian tribes within the restricted Indian territory some five hundred miles to the southward seemed temporarily at peace. From 1873 to 1876 the attention of the army centered upon the Sioux and their allies. Gradually, after Custer had cleared the Black Hills region, the Indians were pushed westward into Montana.

Sitting Bull and Crazy Horse had proved themselves redoubtable leaders, even against the superior forces, equipment, organization and discipline of the Federal troops. They were to have one more moment of glory.

General Sheridan, in command of the troops, planned a final decisive blow. He ordered three divisions under Crook, Terry and Gibbon to crush Sitting Bull's forces on the Yellowstone. Crook encountered and defeated Crazy Horse's band, but Terry and Gibbon failed to find the other Indian forces. The Indian warriors to the number of about six thousand assembled again on the Little Big Horn.

Custer was ordered with a cavalry regiment of six hundred men to bar an eastward movement of the Indians until the other troops could reach the scene. Many and violent have been the arguments as to the precise responsibility for what happened. The consensus appears to be that Custer, acting upon erroneous information that there were only some 1,200 or 1,500 Pawnees in the Indian encampment on the march to join the Sioux under Sitting Bull, decided to strike instead of waiting for re-enforcements as ordered.

The statement that Custer acted on erroneous information has been disputed by no less an authority than his own chief of Indian scouts, a Crow Indian known as Half-yellow-face. The testimony is recorded by Frank B. Linderman in *American,* the biography of the most famous of Crow Indians, Chief Plenty Coups, published by the John Day Company. Linderman lived for more than forty years as a friend and student of Crow, Cree, Chippewa and other Indian tribes at Flathead Lake, Montana. His word on those Indians is authoritative.

According to his account, obtained directly from Chief Plenty Coups, Half-yellow-face on the morning of the battle warned Custer against dividing his forces, saying that there were too many of the enemy to be attacked successfully even if the entire Custer force was held together. In reply Custer is quoted as saying, "You do the scouting and I will attend to the fighting."

In any event, Custer made the error of dividing his forces with the idea of surrounding and surprising the Indians. Custer himself with 260 men attacked what he believed to be the center of the Indian force. And quickly he found his own force surrounded. The slaughter was terrific. The last chance of the troopers vanished when, instead of the detachment under Major Reno which they had expected to attack the savages from the other side, a fresh band of one thousand Cheyennes came whooping into the fray. Every man in Custer's division was killed. The Indians themselves lost only forty-two dead.

Despite the defeat and death of Custer, apparently there was no loss of respect among the Crows for the glory-hunting general. The Crows had been traditional friends of the whites for many years, as they had been traditional enemies of Cheyennes and other plains tribes. Wiser than many of their savage neighbors, they had tried to retain their lands by favor rather than by force. They had rendered notable service as scouts for U. S. troopers in numerous campaigns. They admired the dandified and reckless Custer.

Lieutenant James Bradley, who was with General Terry when the latter moved to the support of Reno's detachment of the Custer regiment after the massacre, has left a vivid description of the mourning of the Crows when news of Custer's death reached them. "One by one," said Bradley, as recounted by Linderman, "they broke off from the group of listeners, going aside a little distance to sit down alone, weeping and chanting their dreadful Mourning Song, and rocking their bodies to and fro."

The Custer Massacre had been a major catastrophe in frontier warfare with the Indians. But the subsequent drive which brought about the surrender of remnants of the various tribes cleared the way to the end of that frontier as a frontier.

Sitting Bull fled with large numbers of his people into Canada, but returned eventually under amnesty offered by General Nelson A. Miles. In the year following the Custer Massacre the Nez Perce tribe in Idaho, despite the inspired leadership of their famous Chief Joseph, was reduced to impotence in a campaign of destruction and perfidy which is one of the blackest marks in all the history of the white man's dealings with the red.

Sadly the remnants of the Nez Perces submitted to forced removal into a reservation in Indian territory. Later, after hundreds had died of illness in a lowland climate to which they were unaccustomed, a belated consideration. of their tragic unhappiness permitted them to return to a reservation in Idaho. Blackfeet, Flatheads and others were established on small scattered reservations in the Northwest.

While Sitting Bull and his followers were moving into Canada, a large number of the Northern Cheyennes joined the Sioux under Crazy Horse in a winter camp on Powder River. In the spring of 1877 the Northern Cheyennes surrendered to the troops. Nearly all of them were moved into the reservation of the Southern Cheyennes who had been established with the Arapahoes in the former Seminole lands east of the Texas Panhandle. There they faced new and disastrous conditions.

The climate was hot and humid, in contrast to the high dry mountain air of North Dakota and Montana in which they had lived. Wild game was depleted, in contrast to the natural supply which still existed in their northern homelands. Malarial fever and starvation began to take heavy toll of life. The Federal agent on their reservation testified before a Senate committee that the supplies of government rations were never more than enough to maintain life for more than three-fourths of the year, and the quality was extremely poor. Still the government did nothing about it.

The Northern Cheyennes endured that situation for a year, and then struck out for their homeland, asking no permission and making no concealment. With fewer than one hundred warriors and more than two hundred old men, women and children, they managed to move all the way from the Red River country into Montana, fighting repeatedly against Federal troops which were sent in pursuit or ordered out from various forts and railroad lines which they must pass. The number of Federal troops engaged from time to time has been estimated as high as thirteen thousand.

In its way it was an epic accomplishment, an astonishing demonstration of the stamina of the Indians and the power of their love for their wild homelands. That three hundred-odd Indians could travel across nearly a thousand miles of country, taking their living largely from the wild game they encountered, and defeating or dodging thousands of trained, rationed, armed and mounted soldiers, was an heroic accomplishment.

Chief Joseph with his Nez Perces had made a somewhat similar flight from the troops with a larger group through Idaho, Wyoming and Montana country, but handicapped by far more women and children, in the previous year. That story is another classic of Indian history which has been more widely publicized than the flight of the Northern Cheyennes. Both movements, however, resulted in the same sad denouement. The Indians were trapped and forced to surrender. General Nelson Miles induced the Cheyenne warriors to join him as fighting scouts to clean up the last of the warring Sioux.

A century after the Declaration of Independence, the American Indian was subjugated. Railroads were shouldering the last of the buffalo from the western plains. Steamboats plied the Mississippi, Missouri and Ohio, and many of the lesser rivers. The South had been "reconstructed." The panic of 1873 and its subsequent depression had been weathered, with new hope growing where old hope had met destruction. The shrill yells of the cowboys on the long trails from Texas to shipping points at railheads in Kansas had succeeded the war whoops of savages.

The Far West, the Northwest and the prairies had been settled and made safe. With the outstanding exceptions of paved highways, automobiles, airplanes, moving pictures and radio, the United States was much as it is today. Mark Twain had written the immortal *Adventures of Tom Sawyer*. The great Sarah Bernhardt was preparing for her first successful tour of the United States. The theaters of New York and San Francisco and a score of cities between were gathering places of beauty and culture rivaling that of the Old World. Thomas Cole, famed artist, had long since painted his famous group called "The Course of Empire," revealing majestically the sway of civilization from savagery to magnificence and on to the desolation wrought by man's inhumanity to man. Thomas Edison was enlightening America with practical science. Horace Greeley's famous dictum, "Go west, young man," was already history which had proved its justification.

The West had been conquered, settled, and civilized. The frontier which had been inviting adventurers and pioneers for centuries was at last restricted to an area of perhaps one-fiftieth of the United States—the area of Indian occupation within what is now Oklahoma. The states of Missouri and Arkansas on the east, Kansas on the north and Texas on the south and west were its boundaries. Indian territory was not even a territory as a political entity within the Union, as New Mexico, Arizona, Wyoming and others were territories, with governors, competent courts and representation in Congress. It was merely the home of the Indian, restricted against settlement by whites, governed largely by its various tribal councils.

This, little more than half a century ago, was the last frontier. There, within an area approximately 350 miles long by 200 miles wide, were gathered nearly all the types of men, living under much the same conditions which had made the frontiers of America lands of opportunity and hardship through three centuries preceding. While the rest of the United States continued on its way of growth, prosperity and civilization, Indian territory for the moment stood still.

Pointing up the highlights of the history of the United States as in a pageant, that was a dramatic moment. Latent forces, unrealized by man or man's government, impended.

CHAPTER IV

CATTLE DRIVES AND BOOMERS

FATE, as revealed first in the secession of Texas from Mexico and later in the War Between the States, was preparing to make the Indian territory into America's last frontier while the flood of migration and the subjugation of the savages on other frontiers were doing their part. Texas was already becoming famous for its cattle when that Republic joined the Union of the States.

Through western Oklahoma, Kansas, Nebraska, Dakota, Wyoming and Montana the buffalo constituted the most important resource of the Indians. It provided their most reliable food supply and the all-important shelter of their teepee homes. Herds sometimes extending beyond the reach of man's eyesight on the level plains moved northward from the Red River country with the spring and southward from the Canadian prairies before the winter snows. But the buffalo was as helpless against the competition of man-herded cattle as were the savages against the white men.

The beginning of the end of Indian power in the West was simultaneous with the beginning of the end of the buffalo. The white men, whose insatiable demand for furs had stripped the beaver streams and the natural supply of mink, otter and other pelts within a century, required only a few years to destroy the buffalo. To the first settlers, even to the first builders of the Union Pacific Railroad pushing out across Iowa and Nebraska, that had seemed impossible.

Those countless thousands of shaggy-shouldered beasts

39

seemed beyond the power of man to destroy. The Indians had lived upon them for centuries, and still they had increased. They impeded the wagon trains of early emigrants, and were butchered without mercy. Still the great herds survived. They blocked the railroad trains of an advancing civilization. They ruined the first tilled farms. They seemed as persistent and implacable as the pioneers themselves.

But when a promise of profit from their slaughter appeared, when a market was found, not for buffalo meat but for buffalo robes, in the eastern cities and countryside, the herds were doomed. A thousand skilled huntsmen moved to the attack. A single sharpshooter could destroy fifty buffaloes in a day. Hundreds did so, and repeated the slaughter day after day as the herd grazed onward. Whole trainloads of hides moved back to eastern markets. A few short seasons of that, and the great herds were no more. The ranges were cleared for cattle, and the plow.

Texas had the cattle, a heritage from the Mexicans. At the outbreak of the Civil War more than four million longhorns roamed the Texas ranges. The market for Texas beef within the Confederacy was cut off with the capture of Memphis, Vicksburg and New Orleans. When the war ended Texans returned to their ranches to find six million longhorns, quoted at an average of two dollars a head, although beef was bringing twenty-five cents a pound in New York.

Stock within reach of gulf ports might be shipped to eastern markets by sea, but the bottoms were few and the cost high. A new railroad was building westward into Kansas. The Indian peril had been diminished. A broad and grass-grown way extended from Texas through Indian territory toward the railway.

Jesse Chisholm, son of a Cherokee mother and a Scottish father, scout, guide, trader and pathfinder, had traveled much of that way in 1865, along a route now approximated by the Rock Island Railroad, from Wichita, Kansas, to the Indian agency at the site of the present Anadarko, Oklahoma. Chisholm had approximately retraced the route followed by Federal troops retreating from frontier posts in Confederate territory four years earlier. That journey had been made under the guidance of a Delaware Indian named Black Beaver, but the trail which was to become one of the highways of history in America gained its fame as the Chisholm Trail.

The Chisholm Trail beckoned to the cattlemen of Texas. When Joseph McCoy of Chicago arranged with the extending railroad for favorable freight rates on cattle from the end of the line at what is now Abilene, Kansas, the historic drives began. McCoy spread the word through Texas. The cattle moved from distances sometimes as great as one thousand miles. The grazing was excellent. The roots of the grass which had supported the buffalo herds for centuries still held the soil. White blizzards occasionally caught the trail herds, but there were no black blizzards of dust in those days.

It was a long and arduous journey, but the feed was free, the Indians generally inoffensive, the cowboys tough and the market profitable. As the railroads pushed westward, the shipping points moved with them, resulting in approximately parallel but somewhat shorter trails. Abilene was the first great shipping point, Dodge City the last. For a time Dodge City was the cowboy capital of the world, and perhaps the toughest town.

A thousand tales of which the fictional could not better the facts have been written of those cattle drives. In a score of years 15,000,000 head of cattle moved across the Indian territory. Usually a dozen men were required to manage 2,500 longhorns, the most practical average size of a trail herd. A trail boss was in charge, with nine cowboys, a horse-wrangler, and a cook who drove the chuck wagon. Such a group, with grass and weather conditions favorable, could move 2,500 cattle from the Texas ranges to the railroad at a cost of from forty to eighty cents a head, depending upon the length of the journey.

Frequently conditions were not favorable. Thunderstorms, wolves, a remnant of the dwindling buffalo herds, or a few wily Indians might start a stampede or otherwise disturb the drive. A stampede could run many thousands of pounds of beef off a well-fed trail herd in a few hours. It could delay progress for days while the scattered animals were rounded up.

It was a tough, hard life, in which absolute freedom was the one great advantage, outweighing all disadvantages. It was chief of the factors conspiring to restrict the Indians far more closely than they had already been restricted.

As has been mentioned, the Cherokees under the treaty of 1866 had accorded the government permission to settle so-called friendly tribes in their broad western Outlet to the buffalo country. The settlement of those tribes and the destruction of the buffalo herds had extinguished the original purpose of the Outlet, but had not extinguished the Cherokee title to a vast area of land. It was thus perhaps that the Outlet came to be known as the Cherokee Strip. It was no longer an outlet.

It was still untilled, unmarked by any towns other than Indian villages. It was rich in buffalo grass, and a way to profit for the Texas cattlemen. Delays upon the long trail frequently were turned to greater profit by the owners who fattened their stock by the wayside. When beef prices were low, or delays in transportation made it advisable, the herds grazed the Strip for months at a time. The Texans were wise hombres, long trained to take full advantage of any opportunities which the frontier offered. It was inevitable that they should establish supplementary ranches on the western prairies of the territory. The Strip was dotted with cabins, dugouts and corrals, and the longhorns upon it became as numerous as the buffalo had been before them.

But the Cherokees were civilized. They understood their rights. In 1880 the Cherokee Council voted to levy a head tax of one dollar each on all cattle within the Strip. Shocked and chagrined were the cattlemen. Their protest was prompt and emphatic, and partly effective for the moment. The tax was reduced to forty cents for two-year-olds. And the cattlemen paid—but upon their own reports of the numbers grazing. For their further protection, in the three years following "The Cherokee Strip Livestock Association" took definite form, with bylaws, brand registrations, elected officers and dues.

The headtax system had not proved satisfactory. The Association treated with the Cherokee Council and leased the entire unoccupied part of the Strip for five years at $100,000 a year, subleasing to the cattle owners in its membership. The great spring and autumn roundups, in which hundreds of cowboys representing scores of ranch brands gathered and sorted hundreds of thousands of scattered animals, became high points of life on the plains.

But frequently, especially after a severe winter had driven the stock many miles from their home range, the roundups were slow and expensive. The peculiar notion of the white man in contrast to the Indian that land should be defined and owned by the individual brought fencing upon the prairies. The drifting of the cattle before the winter storms, their natural habit and protection, was checked. And the god of Nature, perhaps the god of the Indians, again revealed the error of the white man's way.

The blizzards which swept the Cherokee Strip in the winter of 1884-5 destroyed almost the entire herd of numerous ranches. Charles F. Colcord, whose noteworthy career from boyhood on a Texas ranch to wealth and leadership in Oklahoma will be told in its place, counted several thousand head of cattle standing in a foot of snow, packed in a fenced angle of a single ranch, frozen to death. Ruined cattlemen, seeking to appraise their losses, could walk in places for a mile upon the frozen bodies of the animals.

But the persistence of the pioneers was irresistible. At the end of the original five-year lease the Association increased its offer from $100,000 to $200,000 a year and renewed its agreement with the Cherokees for another five years. The white man's foot was in the door. Never in the history of America has it been permanently withdrawn. In the Cherokee Strip it was the foot of the cattleman. In the so-called Unassigned Lands to the southward, purchased by the government from the Creeks and Seminoles in 1866, it was the foot of the so-called boomer, under the leadership of David L. Payne.

Payne was of the kin of Davy Crockett, of the blood and fiber of the pioneers of America. Born in Indiana in 1836, he had moved on with the frontier until at the age of twenty-one he was establishing a homestead in the untilled lands of eastern Kansas.

Little more lettered than Davy Crockett himself, he had risen to a seat in the Kansas legislature, served three years in the Union army, led a company under Custer in one campaign against the Indians. Unlike Crockett in friendship and advocacy of justice for the Indians, Payne held the more common view of the frontiersman that Indians were Indians, and nothing more.

In his campaigns under Custer he had become familiar with the Unassigned Lands along the North Canadian River. He coveted the land, of which large sections were still unoccupied. But they were within the territory which the government had designated as Indian, and in which white men were barred from settlement. The fact that that was an arrangement of potential advantage to the Indians made it seem no less unfair to Payne.

He contended that both the Unassigned Lands to the south and the lands of the Strip itself were open to settlement under the homestead law. His efforts served to reverse the popular dialogue: "You can't put me in jail for that." "But you're in." Payne maintained that the Federal troops stationed within the territory to maintain the status quo of the Indians could not put him out, but they did.

After the persistent Payne and his various companies of boomers had been thrown out of the Unassigned Lands two or three times, he planned another venture which could at least provide grounds for a test case and decision in a Federal court.

In February, 1883, Payne started from Arkansas City, Kansas, with 132 wagons, 553 men and 3 women to establish a settlement on the North Canadian in the vicinity of what is now Oklahoma City. Payne's ace in the hole was a man named Ackerley, who was to sell whiskey to the Indians in the region which they sought to homestead. It was planned that Ackerley's arrest and trial on the whiskey-selling charge would establish the fact that this area was not Indian land.

It was a difficult journey of some two hundred miles over the winding, poorly marked wagon trails, in weather so cold that the heaviest wagons were able to cross the Salt Fork River on the ice. And within a day after they had pitched their far-flung camp upon the chosen site, a company of soldiers from Fort Reno, forty miles away, arrived to throw them out.

Payne, W. H. Osborn, secretary of the proposed colony, and Ackerley were arrested and placed under guard. The remaining 550 boomers were warned that anyone found in camp the next day would also be arrested. When day broke all but one, C. P. Wickmiller, were on their way back to Kansas. Wickmiller was taken into custody with Payne, Osborn and Ackerley, and the four were marched the forty miles to Fort Reno and jailed to await trial. Payne's plan to make a test case of Ackerley's arrest netted him precisely nothing. The government declined to prosecute. Soon the four men were released and returned to Kansas.

That was Payne's last effort to establish a settlement in the Unassigned Lands. A final effort was made in the Cherokee Strip in 1884. The fact that 1500 boomers could be gathered from a sparsely settled area for that attempted land-grab emphasizes the eagerness of American frontiersmen for free homesteads only a trifle more than half a century ago. Payne led his boomers a few miles south of Caldwell, Kansas, to an encampment on Chikaskia River, called the Rock Falls colony.

There, in violation of a definite ruling from the Secretary of the Interior, chief authority over Indian lands, Payne established a newspaper, *The Oklahoma Chief,* first newspaper to be published in the Cherokee Strip. The first issue appeared in May. By August the demand for the four-page papers at ten cents each had grown so great that the final issue, printed in August, kept the old Washington hand press operating continuously for fifty-odd hours.

The government could not overlook such flagrant violation of its orders. The boomer colony was just within the Cherokee Strip line, but it was a foot in the door. A detachment of Negro soldiers under command of Lieutenant Day appeared from Fort Reno. Grant Harris, the youthful printer in charge of production, and six others of the leading colonists were arrested and placed in irons. The remaining 1500 boomers were lined up and headed back into Kansas. It was not a long trip. When young Harris was asked by Lieutenant Day how long it would take him to get back to Kansas if his irons were removed and he was provided with a pony, he said, "About fifteen minutes."

It was done. The soldiers did not take the boomers too seriously. The other, older men under arrest were taken to the Federal court at Fort Smith, Arkansas, where they were released after six weeks. The confiscated printing plant was dumped in the Cimarron River. A short time later Captain Payne dropped dead at the breakfast table amid a group of his leading boomers in Wellington, Kansas.

He had accomplished one thing. His repeated efforts and failures had won recognition as news among many leading metropolitan newspapers. The news had impressed the nation with the fact that a vast and apparently promising agricultural country, owned in large part by the government, was still virtually unsettled by white men. The cattlemen were hardly counted. As yet they had made little claim upon the land. They were content to drive their herds across the Unassigned Lands and to pasture them in the Cherokee Strip by agreement with the Indians.

But farmers and cattlemen have been at odds probably since the beginning of civilization. The Cherokee Strip Livestock Association sensed a threat in Payne's efforts to establish his boomers on the land. The farmers would plow under the buffalo grass, and otherwise restrict the range. While the Association's original lease of grazing rights at $100,000 a year was still in effect, a syndicate including the richest men in the Association made a written offer to Chief Bushyhead of the Cherokees to buy all the grazing lands in the Strip at three dollars an acre. Probably no more ambitious private land deal has ever been conceived. A large part of the nearly nine million acres in the Strip was involved. How the syndicate could have financed the project must remain a question. Circumstances prevented an answer.

A somewhat similar sale of some 450,000 acres which formed the original Cherokee Strip in southern Kansas for one dollar an acre in 1866 had been followed by litigation under which titles were not definitely settled for sixteen years. Neither the Cherokees nor the government wanted a repetition of that litigation and speculation.

The offer of the cattlemen's syndicate was never formally acted upon by the Cherokees. Instead, the grazing lease was renewed at $200,000 a year. And once again the hard hand of the United States government closed upon the Indian. That was half a century ago, a decade after the Custer Massacre had convinced Americans in comfortable homes from New York to San Francisco, owners of prosperous farms throughout the nation and politicians in Washington that votes by white men were more important than justice to Indians.

Congress authorized Federal purchase of the Cherokee Strip lands at $1.25 an acre, a total of millions less than the cattlemen's offer. In the face of that difference the sale to the syndicate was blocked. The Cherokees were receiving an income of $200,000 a year from the grazing lease. They were reasonably content for the moment. Not so the politicians. John W. Noble, Secretary of the Interior, final authority under the President, prepared the way to enforce the will of the government by declaring that the Cherokee title was invalid; that it was nothing more than an easement which the Indians had forfeited by failure to use the lands. The fact that they had been cut off from those lands in part by the government's settlement of "friendly tribes" in the way to the hunting grounds was ignored. The Commissioner of Indian Affairs dissented from that opinion, and declared that the Indians held a valid title in fee simple.

But Secretary Noble's authority was the greater. He declared the Cherokee lease of grazing rights to the cattlemen invalid. He ordered the removal of all ranch cattle from the Strip. Having thus put the pressure upon the Indians by cutting off $200,000 a year of their tribal income, he moved on with the traditional consistency of a successful politician. Having declared that the Indians had no title, he took the necessary steps to buy the lands at less than half the price offered by the cattlemen. The Cherokees bowed to a proclamation signed by the Great White Father, President Harrison, and gave up their lands to the government for millions less than they had been offered by private purchasers.

Half a century after their fathers had been moved from their ancestral lands, out upon the trail of tears, by President Andrew Jackson and the commonwealth of Georgia, the Cherokees as a nation found themselves exploited for the last time by the white man's greed. Thus was the way cleared for the advance of civilization.

The last frontier was growing more and more restricted. But before it ceased to be a frontier it was to reveal itself as a setting of adventure, opportunity and crime seldom exceeded in the history of America.

By the 1880's, it should be remembered, the population of the United States had grown, in a century, from three million to ninety million. It had grown in white-populated area from a strip averaging two hundred miles inland along the Atlantic seaboard to an area nearly fifty times as great. Cities, schools, courts, railroads, factories, farms and all the other appurtenances of civilization, including political subdivisions and pressure, were common from coast to coast, except in Indian territory. That alone was still barred to unrestricted white settlement, its people ruled for the most part by tribal councils.

In some sections it tolerated the cattlemen, who were a hardy lot, and in some sections it offered a haven to desperadoes who were tougher. But more pacific white settlers such as Payne's boomers were invariably driven back outside its borders. That, broadly speaking, was the situation when the Federal court established at Fort Smith, Arkansas, with Judge Isaac C. Parker on the bench began to make itself felt in the affairs of the territory. Judge Parker sentenced 168 men to hang and saw more than half of them executed.

Some indication of the extent of the task which the court expected to perform is revealed by the fact that its gallows was built of sufficient size to hang twelve men at a single fall of the trap. It was never used to its full capacity, but it was ready. On two occasions six men were executed simultaneously. On three occasions five were hanged together. Frequently groups of three or four were dropped to death by a single pull upon the hangman's lever.

The frontier gallows of an earlier day which is still to be seen behind the little courthouse at Downieville, California, could never have competed with that engine of destruction, nor did it need to. The use of the rafters of a house in process of construction in Alder Gulch, Montana, to hang five murderous bandits at one time was a makeshift. That was the action of a group of vigilantes, not of any legal court, to destroy Sheriff Henry Plummer's secret gang of murdering highwaymen. It was never repeated in anything like such wholesale manner. To be sure, the Montana vigilantes did a pretty thorough job, hanging twenty men in a single month in 1864. But all that marked a frontier which had been superseded by civilization for many years before the Fort Smith gallows was built with the sanction of a Federal court.

The Fort Smith gallows had a job before it which was to continue for twenty-one years, until the last frontier itself had ceased to be. The years of its power and effect were simultaneous with the final years of frontier life in America.

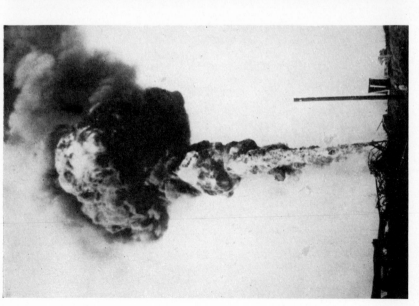

A burning oil well (left) and a burning gas well (right) in the Oklahoma City field. The gas fire, in a Sinclair well, could be seen at night from a distance of fifty miles. It burned for twelve days, consuming 350,000,000 feet of gas a day, until blown out with nitroglycerine.

CHAPTER V

HELL ON THE BORDER

WHEN Judge Parker mounted the Federal bench at Fort Smith in 1875 the court held jurisdiction over seventy-three thousand square miles of Indian territory. Its arm must reach through a wilderness more than three hundred miles to the west and more than one hundred miles to the north and south to seize and punish the criminal.

The difficulties are revealed by the fact that during the years of Judge Parker's incumbency sixty-five United States deputy marshals were murdered by the desperadoes with whom they dealt. Even before the approach of the railroads desperadoes had begun to gather within the territory. The scourings of Nevada, Montana and Colorado mining camps, pushed out of their habitat by vigilantes or by the advance of civilized law and order, had drifted into a region where they expected no restrictions. Renegade cowboys who preferred a life of idleness, debauchery and crime to sixteen hours a day in the saddle on the cattle drives established themselves.

With the coming of the railroads, beginning with the Missouri, Kansas & Texas which entered the territory from the north in 1871, a new type of bad men began to gain a footing within the frontier. Some of these were gamblers and desperadoes with a background of city life and crime. The railroad construction camps witnessed such scenes of violence that the railroad builders were forced to call upon Federal troops to protect them. Some of the criminals were driven back to the cities from whence they came. The worst major criminals who did

51

not dare return to their old haunts, scattered more deeply within the unpoliced areas of the territory.

That was the situation faced by Judge Parker's court at Fort Smith. At the moment the total population of the territory numbered approximately sixty thousand, of which probably fewer than two thousand were white men unconnected with the Indians by blood or marriage. From the latter group Judge Parker arraigned and tried ninety-one prisoners, or approximately five per cent of the white men within his jurisdiction, in the first term of his court. Sixty were convicted. Eight were found guilty of murder, and promptly hanged. And even so the authorities estimated that at least three-fourths of the perpetrators of crime within the year were not even arrested.

Even before Judge Parker undertook that staggering task, an organized criminal band was establishing precedents of gang practices and codes which were to become infamous half a century later in Chicago, New York, and other civilized centers. For example, consider the career of John Childers.

Childers was the son of a white man and a Cherokee woman. He had been born and reared within the Cherokee Nation. By the time he was twenty-one years old he had attained prominence in a criminal band of whites and mixed-blood Indians who lived by violence and rapine. He had already clashed with a U. S. deputy marshal named Vennoy. With a misconceived idea of vengeance he had killed a man who had had nothing to do with that trouble but who happened to hail from Vennoy's state.

Shortly afterward he encountered a peddler named Rayburn Wedding driving a team in which one fine black horse appealed to the eye of the outlaw. When Wedding refused to trade the horse for anything that Childers could offer, the outlaw calmly cut the peddler's throat and threw the body into Caney Creek. Then he stripped the harness from the coveted horse, transferred his saddle from his own animal and rode away. The abandoned wagon and the body of the murdered man were quickly discovered. Scores of persons were able to identify Childers' horse, which had been left at the scene of the crime, and others reported seeing the outlaw riding Wedding's black horse.

Deputy Marshal Vennoy set out with a posse to arrest the murderer, and succeeded with little difficulty. It was the beginning of a series of arrests, escapes, court delays and similar incidents such as were to become painfully familiar in the history of gangsterism half a century later.

Childers, in irons, escaped from the posse. A few weeks later he was again captured, and that time lodged in jail in Kansas. The Federal court then charged with the administration of justice in Indian territory was located at Van Buren, Arkansas. Childers was taken there to await action of the next grand jury. Before the jury could act, he and six other prisoners broke out of the Van Buren jail.

Marshal Vennoy was as irritated as J. Edgar Hoover might be. Casting about for a way to take the murderer into custody a third time, he discovered in Fort Smith a woman of the town who had been intimate with the desperado. He elicited her help on a promise of ten dollars' reward for the capture. Ten dollars! Shades of the Woman in Red! And of various molls of a modern day who have betrayed their gangster-sweethearts.

The woman managed to get word to Childers that she longed to see him before he fled the country. He came, and in the night when his six-guns were laid aside Deputies Vennoy and Peevy entered through a door which had been left unlocked for the purpose, slipped to his bedside and snapped handcuffs upon him. This time he was thrown into the more substantial jail which had been constructed at Fort Smith. The practice of delay which was to grow through the years in the defeat of justice became effective. Counsel obtained postponement of trial until the next term of court, and won the prisoner's release on bail. The gang was all behind him, intimidating witnesses, providing money for lawyers.

Childers felt as certain as any modern public enemy with a battery of unscrupulous attorneys and ample funds from crime that he could "beat the rap," although that was not what it was called in those days. But he was mistaken. The verdict was guilty. There was no appeal from that court except to the mercy of the President of the United States.

The Fort Smith gallows which was to take the lives of eighty-eight men was built. John Childers was the first to die upon it, while two thousand persons, assembled from miles around, with the woman who had betrayed him in the front rank, looked on. With the rope around his neck he confessed the murder of Wedding. In the next instant he established a precedent in the code of gangsterism.

Looking out upon the crowd which had gathered to see him die, he recognized a dozen of his criminal associates whom he said had sworn to aid each other in any and all circumstances. He mentioned the fact to the marshal in charge of the execution. "They don't seem to be doing much for me now," he added.

"Give me their names and I'll give you a reprieve," the marshal offered.

"Didn't you bring me here to hang me?" Childers demanded.

"Yes."

"Then why in hell don't you do it?"

That was the precedent of silence in the face of death, the gang code, set by one of the first famed bandits of Indian territory to die legally by the noose. A moment later his body jerked at the end of the rope.

Only a few weeks later three more killers, all Cherokee Indians who had murdered two white trappers for loot valued at less than twenty dollars, were hanged on the

HELL ON THE BORDER

same gallows at one jerk of the lever. Three others convicted of three separate murders were hanged together the next spring, and another shortly afterward. There seemed no need of a vigilante organization in Indian territory.

At the same time the western part of the territory, farther from the hand of the law, still subject to the vagaries of Indians who had not submitted to white domination so completely as the Five Civilized Tribes, was dangerous ground. A war party of several hundred painted Indians, rising in fierce effort to check the white man's slaughter of the buffaloes, attacked an encampment of hunters at a spot known as Adobe Walls. The whites made a valiant defense with their long-range buffalo guns, but every one was killed. That, incidentally, was the year of the last general uprising of the Cheyennes.

Fear of the western Indians was widespread. Settlers in southern Kansas gathered for mutual protection in various small towns along the line of the new railroad. The frontier was excited and fearful to the verge of panic.

In that situation Pat Hennessey, George Fant and Thomas Calaway ignored the warnings of their friends and set out with a wagon train, freighting supplies from Wichita to the garrison at Fort Sill. At the stage station in the Cherokee Strip near the present town of Bison the freighters again were warned of grave danger. But Hennessey was an experienced freighter, somewhat contemptuous of the Indians. Within six miles the party was attacked and massacred.

Hennessey's body was found spread-eagled between the wheels of his wagon, fire still smouldering around him and his legs almost burned away. The bodies of Fant and Calaway lay near. Later there arose considerable discussion as to whether Hennessey was murdered by Indians or by white desperadoes who were already infesting the region. Preponderance of evidence pointed to the Indians. No one was ever punished for the crime.

Back in the eastern section of the territory, U. S. marshals and the U. S. court were doing better in their efforts to combat the crime of the frontier. Six men were arrested in six weeks and convicted of six murders in the summer of 1875. All six were hanged with one drop of the gallows trap only two months later. Word of such prompt and final justice traveled swiftly throughout Indian territory. It did not stop crime, but it made the criminals wary.

However far they fled, the shadow of the Fort Smith gallows and the horrors of the Fort Smith jail pursued them, and at times caught them. As far away as what is now the Oklahoma Panhandle, then known as No Man's Land, a signboard stood at a trail-crossing pointing eastward, with the legend "Fort Smith 500 Miles."

So far that, in the autumn of 1882, when the certainty of justice in Judge Parker's court had been well impressed throughout the territory, six desperadoes, fleeing from a posse, checked their horses long enough to riddle the "Fort Smith" part of the guidepost with bullets. It was a gesture of defiance and assumed contempt. But the posse returning over the same trail a few days later, after the outlaws had scattered and escaped, discovered an impressive further alteration of the sign. Words apparently carved with a dirk in the hand of a mounted man had improved the sign to read "500 Miles to Hell." And many made that journey within the next few years.

Outstanding among the bandits whose names survived the total eclipse of violent death was one Ben Cravens. In this modern day perhaps his greatest claim to fame may be based upon the fact that it was he who established a precedent to the late John Dillinger's escape from an Indiana jail by the use of a wooden pistol.

Cravens had a long and checkered career, beginning as a bootlegger in Kansas and the Cherokee Strip. From that he moved to cattle rustling in the Osage country, where he was soon arrested. He promptly broke jail in the little town of Perry and freed several other prisoners. His organized gang started immediately upon a career of bank, postoffice and country store robberies. Wounded in one of those robberies, he was sentenced to fifteen years in Lansing penitentiary, where he was assigned to the coal-mining gang. Working beside him, several hundred feet under ground, was Joe Ezell, a notorious bad man.

The two convicts managed to fashion wooden replicas of the long-barreled revolvers popular at the time, and covered them smoothly with tinfoil to give the appearance of the nickel-plated six-guns. With these wooden guns they held up the guard at the bottom of the shaft, stripped him of his own weapons and forced him to signal for the hoist. At the top of the shaft they escaped at a run through a fusillade of bullets from the surface guards. Neither convict was scratched.

Back in the territory, Cravens again turned to robbery, with a holdup of a postoffice and store at Red Rock in the course of which Alva Bateman, assistant postmaster, was killed. Bert Welty, a young farmer whom Cravens had enticed to aid him in the robbery, carried the loot. As soon as they were on their horses and momentarily out of danger Cravens turned a shotgun into Welty's face and pulled the trigger. Welty fell without a sound. Cravens appropriated the loot and rode on.

But a few hours later Welty recovered consciousness under a pouring rain and staggered ten miles to the farm of a man named Hetherington which had been one of Cravens' hide-outs. They were physically as well as immorally tough men on the border. Welty recovered, only to be sentenced to life imprisonment for the murder of Bateman.

Before that had been accomplished a posse had surrounded the house of a farmer named Cunningham in the Pawnee country in the belief that Cravens was there. Cunningham denied it. Almost at the instant of the denial

Cravens pushed open the farmhouse door with the muzzle of his rifle, fired a bullet into Tom Johnson, a deputy sheriff, and started a sprint of eighty yards across a plowed field to cover. Sheriff John Chrisman and deputy Joe Weariman, both reputed to be excellent rifle shots, emptied their Winchesters at the fugitive. He escaped without a scratch. Tom Johnson died of his wound.

Cravens simply vanished. For several years he was the most celebrated outlaw of the territory. Almost every unsolved crime within a radius of a hundred miles was ascribed to him, but the man himself was never seen. Years later it transpired that he had not been in the territory during any of those crimes.

Under the name of Charley Maust he had found work on a farm in Missouri, had married and apparently reformed. But when Charley Maust was sentenced to the Missouri penitentiary at Jefferson City on a horse-stealing charge, a fellow prisoner who had been with him in the Lansing prison identified him to the authorities as Ben Cravens. Bertillon measurements of Cravens on file in Lansing were produced, and the identification was legally verified. Conviction of the murder of Bateman, for which Welty was already serving a life term, followed quickly. Cravens was sentenced to life in the Federal penitentiary at Leavenworth, Kansas. Al Jennings, a reformed bandit of the territory, defended the prisoner. Welty was brought from prison to testify. The prisoner, still insisting that he was Maust, not Cravens, was sent to his punishment.

Few careers of crime have been so close in precedent to the vicious career of John Dillinger—starting as a bootlegger, twice breaking jail, once with the help of a wooden gun, killing at least two men and wounding others and at last betrayed by a former friend, though not by a woman as in the case of Dillinger.

The final bloody end of Jim French offers another example of the complete depravity and viciousness as well as the remarkable physical stamina of the frontier outlaws. French had been a member of the Belle Starr gang, and later of the notorious Bill Cook gang. With a single associate he raided the general store of the tiny town of Catoosa in the Cherokee Nation. It was night. The store was locked. French and his associate forced two citizens to break in a window for their entry.

Within, Colonel Irwin, manager of the store, was seated in a small private office which was also used as a bedroom. French turned his rifle upon the manager through an interior window which looked into the store, but at the same instant caught sight of a double-barreled shotgun in the hands of a watchman seated within the store. Before French could swing his gun toward the watchman his companion fired. The bullet split the wall beside the watchman's head and was answered by the blast of the shotgun, which literally tore the top from the bandit's head.

At the sound Irwin had leaped to the office door and jerked it open, with himself behind it. French poked his gun through the interior office window. A second blast of the watchman's shotgun, aimed at French, buried itself in the window ledge just as French's forty-five caliber bullet tore through the body of Irwin. With his last shot gone the watchman was helpless. French forced him to drag in the dead bandit with the face shot away and brains and blood oozing over the floor. Then, moved by some unexplained impulse, he ordered the watchman to place the mortally wounded Irwin on the bed. That done, he turned upon the watchman.

"Now, you coyote, I'm going to kill you," he said calmly after he had inspected the ghastly body of his companion. Slowly he raised his heavy revolver. But the gun that roared when the watchman was within a split second of death was not the gun in the hands of Jim French. Colonel Irwin, upon the bed, mortally wounded, had managed to draw another forty-five from beneath the pillow. Two slugs fired by him passed through the neck of the desperado. French dropped his guns and bolted from the building. The watchman and a few citizens of the town were at his heels. A few minutes later he was cornered in an Indian hut near the edge of town. The watchman gave him the *coup de grace*.

Belle Starr's place in the history of frontier banditry is unique. No so-called gun-moll of a modern day is likely to approach her record. The most the bad girls of today can expect to do is to beguile their gangster friends at leisure, perhaps act as a lookout or drive a car for some robbery, and at last betray the criminal for what seems a suitable reward.

Belle Starr was a bandit and a leader of bandits in her own right. She was no product of slum or depravity, as study of most of the gangsters' girl-friends of today has revealed them to be. She was a woman of breeding, culture and education. Books vied with horses and the love of money as the great interests of her spectacular career. Banditry, murder, and general depravity were the natural and inevitable results of the reckless life which she chose and the companionships which it entailed.

The record of her life and character might qualify her in many ways as the prototype of a fictional Georgia belle more recently famous. Only daughter of Judge John Shirley and Eliza Shirley, well-to-do residents of Carthage, Missouri, of the best Southern antecedents, she was christened Myra Belle Shirley. The name of Belle Starr came with her later fame and infamy. A twin brother, Ed, was a typical gay and reckless youth of the South at the outbreak of the War Between the States.

Belle's first dramatic adventure, the inception of her fame, was through an incident linked with her reckless brother. At the time the twins were only sixteen years old. Ed Shirley had broken home restraints in the first year of the war. Because of his wild-riding, straight-shooting courage he had already become a captain of guerillas under the famous Quantrell. He was the scourge of the Union forces operating in the neighborhood of Carthage. His sister was an unofficial scout through the region, making long rides alone through the countryside and carrying information to the guerillas.

On one such scouting expedition she was intercepted by a Major Enos in the village of Newtonia, Missouri, where Enos was quartered with a troop of Union cavalry. At the moment Enos had sent a detachment to Carthage to capture young Ed Shirley who was reported to be visiting at his home. The girl was taken into custody to prevent her from carrying a warning to her brother. Major Enos, a former resident of Carthage, knew both her and her brother well. He understood her loyalty to her twin and to the Confederate cause. He appreciated her physical stamina and hard-riding ability. Therefore he took her to his headquarters and guarded her himself.

The girl was furious, but her protests were futile. The major simply grinned at her outbursts. She forgot her gentle rearing and cursed with the abandon of the guerillas who had been her friends. Major Enos smiled. She lapsed into sullen silence. Finally, when he estimated that his detachment of troopers had had sufficient start, he released her with a grin and a parting remark.

"Well, my little lady, you can go now. We will have your brother in custody before you can warn him. So sorry to have been forced to detain and annoy you."

The girl wiped tears of rage from her eyes and ran from the room. Her horse stood before the door where her captors had left it. "I'll beat them yet," she shouted as she mounted, whirled the horse and sped straight across country, leaping fences and ditches on the straightest line possible to Carthage, thirty-five miles away.

"I'll be damned," said the major. "She's a real guerilla. I'm afraid I released her too soon. My own men don't ride like that."

It was true. When the Union troopers trotted into Carthage at sundown they were greeted by a smiling girl on a fresh horse. She could not restrain her gloating triumph.

"Are you looking for Captain Shirley?" she asked. "You are half an hour late. He had business elsewhere. He asked me to give you his compliments with regret that he could not remain to receive you as damn' Yanks should be received."

She nodded again to the crestfallen troopers and rode away, as bright-eyed and unwearied as if she had not just completed a thirty-five mile cross-country run. Few of the Union cavalrymen and none of Quantrell's reckless raiders in all that area of the war zone failed to hear of the girl's accomplishment. Her fame was assured, as her taste for daring adventure was stimulated. The fact that her brother had been saved from arrest only to be killed in battle with a detachment of Federal cavalry a few days later merely served to intensify her bitterness and recklessness.

Shortly after the close of the war, when she was hardly past twenty years old, Myra Belle Shirley married one James Reed. Within a year after the birth of a daughter Jim Reed killed a man who had killed his brother. That crime indirectly started Belle upon her career of banditry.

Reed fled from Missouri with his wife and infant daughter to the little town of Los Angeles, California. There a second child, a boy, was born. The couple lived for two years in Los Angeles, but the making of a living was difficult. Jim Reed, with a price upon his head, moved to Indian territory, as many outlaws were doing at that time. Belle followed soon with the two babies.

Reed established a rendezvous at the home of one Tom

Starr, a notoriously bad Cherokee Indian living with a half-breed wife about eighty miles west of Fort Smith. There Belle made a temporary home and was frequently visited by her husband. There she met Tom Starr's son, Sam, several years her junior. In the summer of 1875 Reed was killed by a fellow desperado who sought the reward which had been offered for him, dead or alive. Belle was a widow with two children at the age of twenty-nine.

But she had money and property, including a farm on which she raised blooded horses. Most of her adult years had been spent upon the frontier, in the society of wild and reckless men. She had become as wild and reckless as the worst. But she also had a taste for some of the refinements of civilization. To gratify that taste she established a home and breeding stud in Dallas, Texas. There she maintained what amounted almost to a salon for the entertainment of various admirers, including numerous noted horsemen of northeastern Texas.

In that period she found opportunity to realize what she may have suspected earlier—the practical value of her charms. A tactical error in adding to her stud brought about her arrest on a charge of horse stealing. Locked in the Dallas jail, awaiting trial, she demonstrated that love laughs at locksmiths. A turnkey succumbed to her allurements, released her and fled with her.

From that time on her position as a female desperado and bandit was established. Within the next few years several notorious characters became known as her lovers. Among them was the Jim French whose career has been noted earlier in this chapter. Others were Jack Spaniard and a man generally known under the alias of "Blue Duck."

"Blue Duck" provided opportunity and incentive for one of Belle's most spectacular exploits. The two were "scouting," as the outlaw gentry termed their activities, in what is now the Oklahoma Panhandle. Dodge City,

Kansas, was then the rendezvous of more professional gamblers, gunmen and wild cowboys spending their season's wages after the cattle drives than any other town in the country.

The attraction to "Blue Duck" was irresistible. Evidently he enjoyed not only the temporary affection of his mistress but her confidence. In any event, he borrowed $2,000 in cash from her and journeyed alone to Dodge City, where he promptly lost the money to a group of professional gamblers in one of the town's most notorious dens. When he returned to Belle and confessed his loss she mounted her horse, rode into Dodge City, climbed the stairway to the gambling rooms above a saloon, and covered the gamblers with two pistols. When their hands were in the air she swept all the money from the main table into a bag and backed to the door. A subsequent check-up indicated that $7,000 had gone into the bag.

At the door Belle smiled her most delightful smile. "Gentlemen," she said, "perhaps there is a little change coming to you. You may have it if you will call upon me in the territory. I haven't time to count it out to you at this moment."

With that exploit adding to her fame, Belle extended her field of operations into Nebraska. In 1880 she married Sam Starr whom she had met when her first husband was using the Starr place as a rendezvous. That marriage gave her rights to take up land as a Cherokee citizen, and with her husband she filed on a magnificent claim of timber and bottom land along the Canadian River. A great log house, barns and outbuildings were constructed. There were no roads. Bridle paths led from the retreat through the forests and hills to civilization. Belle Starr for the moment had had enough of the wilder life of the frontier. Her little girl, whom she idolized, was twelve years old. Her son was ten. It seemed an excellent time and an excellent place in which to embrace respectability.

But the friendships and habits of years could not be entirely denied. From time to time outlaws with whom she or her husband or lovers had scouted came riding on sweat-stained horses down the trails to the retreat, in flight from the law. There was nothing to do but take them in. From time to time she revolted at the coarseness around her. Then she packed her bags with silks and furs and silver toilet articles and made her way to eastern cities or resorts where she lived at the best hotels, spent money lavishly, enjoyed the theaters, and stocked up with books for her return to the wilds.

It was noted by peace officers that temporary lulls in the crime waves of the frontier usually coincided with belated rumors of the appearance of an expensively dressed woman, sometimes with two children, living a life of luxury in St. Louis, Chicago, or some noted watering place of the day. But the reports always reached the officers after the woman had again disappeared from civilization. Telegraphic communication was not so common then. Trains were slower.

In 1882, however, Belle and Sam Starr were convicted of some minor crime and sentenced to serve nine months in Detroit. Belle won the approval and consideration of the warden but was unable to go quite so far as she had in her escape from the Dallas bastille several years earlier. She and her husband served their time.

In the meantime her son, Ed, twelve years old, reverted to the type of his father, Jim Reed, as well as his mother, and ran away from the home of his grandparents in Missouri. He awaited his mother when she returned to her home on the Canadian. For a time Belle became a quiet, law-abiding woman, devoting herself to her children.

But only for a time. Her love of horses again asserted itself. She not only bought them and bred them; she stole them. Her daughter, Pearl, was sixteen years old when Belle found herself again wanted as a horse thief. Before a warrant could be served she fled to Arkansas. In her absence her husband, Sam Starr, with a group of Belle's associates in crime, was accused of a postoffice robbery, and surrounded by a posse. The bandits gave battle. Sam Starr's horse was shot dead under him and Sam was wounded in the head. Four members of the posse were left to guard Sam and another wounded man while the other deputies moved away with their prisoners. Sam was believed to be mortally wounded. Probably for that reason he was not very closely watched.

In the circumstances he managed to steal a deputy's rifle, and hold up and disarm the entire group. Then he escaped. But when Belle was informed of the affair she advised the fugitive to give himself up and obtain release on bail. On the way back to the ranch the party stopped at a place called Whitfield to attend a dance. Among the dancers was a man who boasted that he had been the one who had shot Sam's horse in the battle a month or two earlier. A quarrel quickly developed over that, and in the gunplay Sam was killed. At the age of forty Belle Starr was a widow for the second time by violence.

From that time on through the next three years she made her home at the ranch on the Canadian, never appearing personally in any of the banditry which was common in the territory, but generally believed to be the directing head of many of the operations. Well-known desperadoes, including Jim French and numerous members of the notorious Bill Cook gang, were frequent though unheralded visitors at the Starr homestead. Belle, though twice a widow, was still a gay gal, a popular figure in the wild parties and dances of the countryside.

Returning home on horseback, alone, from one such gathering at which she had danced and flirted with desperadoes and deputy marshals alike, she was followed by a person never identified. Within a mile of her home she was shot from her horse, with a charge of buckshot in the back of her neck. The murderer then fired a second charge into her face as she lay on her back in the mud. Belle Starr died as a score of her followers, associates and admirers died—"with her boots on."

The fact that she was never arrested for a major crime was considered by the hard men and women of the frontier as a measure of her shrewdness and ability as a leader of desperadoes. No one doubted that the large amounts of money that she spent from time to time were the product of the hundreds of robberies which occurred within a radius of a hundred miles from her seat on the Canadian River during the years of her activity. Yet she, by hook and by crook, escaped the more serious penalties until an assassin's gun ended her career on her forty-third birthday. It was February 3, 1889. The year was one of the three most important in the history of that country.

CHAPTER VI

The First Great Land Rush

WHILE neither the morals, the ethics nor the economic situation of residents of Indian territory were improving as the rest of the West was improving, the land itself was exerting an increasing attraction to many thousands of white men. Though Payne's boomers invariably had been driven off the Indian lands by Federal troops, they had succeeded in advertising the attractions of the territory throughout the United States. Increasing pressure was brought to bear upon Congress to open the unoccupied lands to settlement.

The pressure was increased by the organized efforts of various railroad companies. Railroads probably were the most powerful single influence in Washington through the years of their expansion west of the Mississippi. They were rich, powerful, unscrupulous. They retained the shrewdest of lawyers and lobbyists. They controlled political machinery, elected representatives and senators to Congress, elected judges who sustained their claims, and otherwise brought pressure to attain their purposes.

Railroad builders and financiers, exerting their influence as far back as 1866 when Creek, Seminole and other Indian lands were ceded back to the Federal government, obtained an Act of Congress providing that every alternate section of land on both sides of railroad rights of way through Indian territory should be given to the railroads if the Indians consented. Similar arrangements helped to hurry the first transcontinental railroad to completion across Nebraska, Wyoming and Utah. A like plan brought

about the Southern Pacific's domination of the economic
and political situation in California through nearly half a
century.

It was a great prize for the railroad barons. But at the
moment the Indians refused their consent. The government decreed that of the first lines building toward Indian
territory only the first to reach the border, one at the east
and one at the north, should have immediate right of way.
The Missouri, Kansas & Texas won the race from the
north, and completed its line across the territory in 1872.
The Atlantic & Pacific, later the St. Louis & San Francisco,
entered the territory on the east as far as Vinita in the fall
of '71.

The pressure brought upon Congress was unremitting.
The Indians were still reluctant to give up land to the
railroads. But white civilization had been overriding the
moral rights of the Indians for nearly three centuries. This
was no time to stop. After all, the whites had developed a
land which had never supported more than two million
natives into one which gave not only support but comfort
to nearly one hundred million persons. If the aim and
justification of life is to improve the numbers and condition of humanity, the end has justified the means.

In the countless centuries of aboriginal occupation of
America the native Indians had advanced so little in the
art of living that they required many thousands of acres
for the maintenance of a tiny tribal group. A score of
acres of the best land were as much as the average village
could cultivate with their primitive tools, laziness and ignorance. Beyond that they were forced to hunt and fish
over wide areas for their living. Starvation as often as
inter-tribal wars kept down their numbers through the
centuries of their savagery. They were, in economic effect,
no more crowded when one hundred million white men
had pre-empted most of their lands than when Columbus
discovered America. On the whole they lived as well on

their limited reservations as they had upon the entire area of the United States prior to white occupancy.

One great advantage they had lost—their liberty. Their freedom to make war, to scalp, to hunt, to feast and to starve had been destroyed. The fierce pride which liberty had stimulated through countless generations had been shaken. Bowing to the superior forces of civilization with less and less resistance as time went on, they had traded their birthright of liberty and pride for a mess of pottage—for what is now euphoniously known as social security.

The situation had boiled itself down only half a century ago to a point where the Indian had almost ceased to be a factor in American life except as a subject for political profiteering and economic intrigue. They were still sufficiently well organized with their tribal governments and defined reservations to form an obstacle to the final exploitation of their land. Numerous and varied were the efforts to overcome that obstacle.

In 1870 representatives of the Five Civilized Tribes and nine lesser tribes gathered in Okmulgee and drew up a constitution for a proposed territorial government with the same rights and representation in Washington as any other territory. In exchange for such questionable advantages, however, the Indians would sacrifice their tribal integrity and much of their common land. The proposal was rejected. The tribes held out against the railroad invasion, even in defiance of some of their own leaders, for another fifteen years.

But at last the growing influence of the railroad builders, together with a popular demand of vote-wielding whites for more free homesteads, had its effect in Washington. In 1886 Congress opened the territory to continued railroad construction under the right of eminent domain.

The right of eminent domain as decreed by Congress in the Railroad Act of 1886 overcame the necessity of obtaining Indian permission for further railroad construction. Immediately the Santa Fe Railroad extended its line from Arkansas City, Kansas, to Ponca City in the Cherokee Strip. In the following year it connected with the line to Galveston. The Rock Island road built southward from Caldwell, Kansas, to Fort Reno. Four smaller companies went into action in the eastern part of the territory. The Frisco line extended across the length of the territory into Texas. The way was cleared for the Oklahoma land rush.

In January, 1889, the government reached an agreement with the Creek Nation concerning the so-called Unassigned Lands of west central Oklahoma. The Creeks were paid something more than two million dollars for relinquishment of their claims to the area. The news spread with the speed of a prairie fire through all the country west of the Mississippi.

The lure of free land had drawn hundreds of thousands of white settlers into the western plains and prairies. For the most part those settlers were a hardy and generally admirable people, eager to work and to win economic and social independence at the cost of immediate security. Under the force of their hands and their character nearly all the original "Indian country" west of the Mississippi had been developed into sovereign states of the Union within a single generation. Minnesota became a state in 1858, Kansas in 1862, Nebraska in 1867. North and South Dakota, Montana and Wyoming had reached a point of development which brought them into the Union as states in 1889 and 1890.

The restricted Indian territory bounded by Kansas, Arkansas and Texas contained the only vast area of fertile land in the United States which could be opened to homesteaders fifty years ago. It was still a period of unrestricted immigration. Hundreds of thousands of Europeans were pouring into the United States each year. Many sought work and found homes in the slums of New York, Boston,

Philadelphia and Chicago. But many others, peasants with a background of countless generations on the soil, traveled westward to join the pioneers of American birth upon the prairies.

Many of the western pioneers who had succeeded were ambitious for new and greater holdings. Many of those who had failed still cherished the belief that all they needed for success was a free homestead. When President Benjamin Harrison, in March, 1889, proclaimed that from twelve o'clock noon, April 22, the central Oklahoma country would be open to homesteaders, the response was prompt and tremendous. Men and women gathered from all sections of the country, though chiefly from the near-by states.

Covered wagons such as had seldom been seen upon the plains in recent decades moved up to the line. Buckboards with fast teams were prepared for the stampede. Saddle animals with stamina to make a run of many miles across the prairie were groomed and trained to win the prize of free acres. The Santa Fe Railroad, the only one then actually crossing that section of the territory, sold hundreds of tickets and saw thousands of passengers swarm into and over the passenger and freight cars coupled to puffing engines at the border. Never in the history of the United States had there been a scene of similar drama.

Federal troops patrolled the line. For several days they had been busy hunting down and dragging out the impatient "sooners" who had sneaked into the territory and hidden themselves to be ahead of the rush. As the morning of the great day advanced the hilarious waiting throng was highly entertained by the ejectment of those crestfallen cheaters.

At last a bugle sounded. A gun shot reverberated along the line. Horsemen plied whip and spur. Buckboards and wagons careened across the line. Trains gained headway. Men, and some women, sprinted on foot through the dust

Oklahoma City as it appeared about one week after the great stampede to the Unassigned Lands in 1889.

(Courtesy *The Daily Oklahoman*)

and racket. The rush was on. The confusion was beyond description.

Bona fide settlers who wanted farms scattered widely through the countryside. But each tank station or siding along the railroad was a potential townsite. Persons most interested in the possibilities of trade or real-estate speculation swarmed to stake town lots. Guthrie and Oklahoma City had been generally agreed upon as the most promising townsites, although neither boasted a single place of business when the starting gun was fired.

Men and women scrambled from the trains before they ceased rolling at each tank stop, ran headlong across the prairie, drove stakes into the ground and established their claims. The greater part of the throng entered from the northern border, but many made the run from permitted areas in the Chickasaw reservation on the south.

The sun had risen upon a virgin country, virtually uninhabited. It set upon a land peopled by one hundred thousand exhausted but exultant settlers. Each of one hundred thousand individuals had experienced the most exciting day of his or her life. No two experiences had been identical.

One young woman exerted her charms so successfully upon the engineer of one of the trains that he permitted her to ride the cowcatcher of his locomotive, and agreed to slow down at her signal. When she had reached a place which attracted her she gave the signal, and leaped to the ground as the engine slowed. Before the passenger coaches came abreast of her she had torn off a skirt, knotted it to a tree, and was shouting her claim to the throngs upon the car roofs and hanging from the windows. That was merely one incident of thousands.

The subsequent career of that young woman has been forgotten with a multitude of others. Very likely her first crop failed, she sold her homestead for two hundred dollars and returned to Kansas to work out her destiny as a waitress in a Dodge City lunch room. That, approximately, was what happened to many of the stampeders. Some worked their way into fame and fortune.

Charles F. Colcord, for example, well-equipped for the race, familiar with the land through his years in the cattle business during the Indian occupancy, won a claim on a broad and level piece of prairie. He sold it immediately to the first man who offered to buy it, and hurried on to the site of Oklahoma City, swarming with claimants of newly staked town lots. There he encountered an old friend from Kentucky who offered to trade him a lot for his team, wagon and outfit. Colcord was willing to look at the lot.

The two men drove to the end of what was called Reno Street, at the railroad right of way. A carpenter drove the last nail in a board shanty and snapped a padlock on the door as they arrived. It was one of the first half-dozen frame buildings in the town. The Kentuckian looked it over with pride, but he lacked the vision of a Colcord who saw a city of more than one hundred thousand population displacing the scattered tents, shacks and covered wagons which covered the surrounding acres. Colcord agreed to the proposed trade, gave up his team and outfit and received the title to Lot 1, Block 1, the first lot surveyed in Oklahoma City. The key to the new shack accompanied the title.

Others of the stampeders had varying experiences. Many a woman, following in the wake of the first mad rush, drove all afternoon and all night with a wagon containing all her worldly possessions, her children, and a coop of chickens, with a cow led behind. Her husband had moved ahead on horseback or by train, with some vague directions for meeting at a certain turning of some stream or beside a clump of cottonwoods where he planned to be first with his homestead stakes.

Many such families came up with their men folks within a day or a night and settled down to the terrific labor of establishing homes and farms. Others strayed far in their ignorance of the countryside and were lost for days. In numerous instances the husband and father left his newly staked homestead to search for the wanderers, and returned to find it pre-empted by claim jumpers.

The proposed city sites, of course, were the scenes of greatest confusion. Chief of these were Guthrie and Oklahoma City. Town lots in Edmond, Norman, El Reno and Kingfisher were also in demand. Speculators, traders and gamblers moved among the excited throngs, buying this lot for $50, another for $100, or a quarter section for $200. Sometimes they resold in an hour at double their money. Many returned to Kansas or elsewhere on the very trains which had brought them in, but with a profit or loss of hundreds of dollars on their day's experience.

The fires of settlers who had made the rush for farm lands, not for speculation, marked the darkness over thousands of acres of prairie that night. Within a week plows were turning the virgin sod. Rows of tents, shacks and stakes defined the cities. Gamblers, merchants and bootleggers set up their establishments in the centers of population. A few United States soldiers kept a semblance of order. There was no sanitation, no town or county government, no system. But there was hope and freedom.

The Anglo-Saxon instinct for order and security quickly made itself felt in temporary organization of town councils. Government land offices were set up at various centers. Streets were surveyed. Disputes over conflicting claims were settled by adjudication when they had not previously been settled by force. Within a week the worst confusion was past. Postoffices were established, bringing in mail for thousands of persons who stood in line for hours before general delivery windows.

Whole trainloads of lumber, food, plows, hardware and other supplies were unloaded beside the tracks at Guthrie and Oklahoma City, and other stations. The sound of hammers and saws, punctuated by the whistling of locomotives, filled the air which through centuries had never known more than the war whoop of Indians or the howling of wolves and the bellow of buffalo bulls. It was a stirring life, but a life still filled with hardship and labor, trial, disappointment, and only occasional success.

The strong, the shrewd, the persistent, the industrious, the unscrupulous, the lucky, survived and prospered to a greater or lesser degree. The weak, the lazy, the unfortunate, and those who could not establish themselves as parasites upon the frontier society found their way miserably back to the places of their origin.

It has been estimated that one hundred thousand persons made that historic stampede into the first free lands of Oklahoma. The virgin prairie could not support so many. The first crops brought bitter disappointment and a definite threat of starvation. Many men made their way back to their home states to seek work through the winter. Wives and children remained upon the lonely homesteads, caring for livestock, maintaining their claims, upholding their faith in the face of white maurauders, begging or stealing Indians, blizzard, sickness and hunger. To those valiant women must go much of the credit for the establishment of Oklahoma. Other forces, economically greater forces, have contributed beyond measure, but the loyalty and stamina of the pioneer women came first.

No powers of individuals could completely offset the hardships of that frontier. The federal census of the following year put the population of the area at 61,384. Nearly forty per cent of the immigrants had failed, and retreated. But the ground had been broken. Civilization had established itself in a primitive land.

CHAPTER VII

THE LAST STAMPEDE

DESPITE the hardships and failures which attended the first spectacular rush into the Unassigned Lands, there were plenty of frontiersmen, pioneers and adventurers, eager to join the subsequent invasions.

The first settlers, without adequate courts or peace officers, had suffered heavily at the hands of cattle thieves and greater or lesser banditry. The Federal court at Fort Smith was too far away to be consistently effective. Division of jurisdiction by which western portions of the territory were assigned to Federal courts in Kansas and Texas had helped but little. The so-called No Man's Land, approximately the area of the present Oklahoma Panhandle, owned by the government without Indian restrictions, was almost unpoliced.

The courts in Kansas and Texas lacked the practical forcefulness of Judge Isaac Parker at Fort Smith. The result, too frequently, had been such miscarriages of justice as to give the bad men of the western area a feeling that robbery and murder were comparatively safe occupations. Even the Fort Smith court had become less effective in the years immediately prior to the first opening of Oklahoma lands.

In the circumstances the newly settled Oklahoma lands felt the urgent need of some more effective authority than the distant Federal courts. And Oklahoma proper, then covering approximately half of Indian territory, was organized as a territory with congressional approval in

1890. Guthrie was established as its capital. The Panhandle country was included and declared open to settlement.

The movement of settlers into the Panhandle was on a much less exciting scale than the rush of the previous year. Cattlemen who had grazed their herds there for years did their best to discourage settlement for farming. Their arguments were wiser than they knew. A generation later, when the sod of those grazing lands had been plowed and harrowed, the dust of its wind-blown acres was to shadow the forgotten cattle trails over more than one state.

In the meantime the way was prepared for a more orderly reduction of the area of the last frontier by opening and settlement of additional Indian lands. A commission appointed by President Harrison began negotiations with various tribes of the western reservations in the year in which Oklahoma was established as a territory. Several of the tribes were persuaded to accept allotments for each member in lieu of the communal holdings of the reservation. As fast as these arrangements were made, the surplus lands were opened to settlement. Each opening was the occasion of another land rush, but smaller than the first stampede, as the areas opened were much smaller.

In September, 1891, some three-quarters of a million acres formerly included in the Iowa, Pottawatomie, Shawnees, Sac and Fox reservations were added to the settled area of Oklahoma Territory. The following year a considerably larger area released by the Cheyennes and Arapahoes in consideration of individual allotments was opened.

In 1893 the final spectacular drama was enacted in the opening of the so-called Cherokee Strip, together with parts of the Pawnee and Tonkawa reservations. In that single movement more than five and one-half million acres were taken over from the Indians and added to the area of Oklahoma settled by the whites. The white settlement of a small area of the Kickapoo reservation in 1895 and a larger area of Kiowa and Comanche country in the Southwest in 1901 were anti-climax. The rush into the Cherokee Strip in 1893 was the last great spectacle of mass settlement of the last frontier. With that movement the frontier as a land of free and hazardous opportunity, as it had been known in America for nearly three centuries, came to its inevitable end. Only the hardships, the banditry and the attendant color of pioneer accomplishment persisted for a few more years. The wide free lands at last were gone.

President Cleveland and Secretary of the Interior Hoke Smith were given broad discretionary powers by the act of Congress which provided for the opening of the Cherokee Strip to homestead settlement. It was hoped to avoid some of the confusion and trouble which had attended the first stampede.

Counties were defined, certain sections set aside as school lands and for other public purposes, a half section was surveyed for each county seat, and four land offices were established, at Perry, Enid, Alva and Woodward. Filing fees ranging from one dollar to two and one-half dollars an acre, depending upon the agricultural advantages of the land, were demanded to balance the government's outlay of cash to the Indians.

The Strip, however, was to be settled by the horse-race method which had been used in the earlier openings. Secretary Smith evolved an idea that the "sooners" could be eliminated by requiring registration and affidavits of qualification to file on homesteads. For that purpose he established a neutral strip one hundred feet wide just within the Cherokee Strip border. There were nine makeshift offices in that neutral strip, each with three snooty clerks from the Department of the Interior to make out the half-dozen different blanks and affidavits required from prospective settlers. Five days immediately prior to the opening were allowed for the winding of this red tape. One hundred thousand persons besieged the nine offices.

It was mid-September, 1893. The weather was hot and dry. The dust stirred by thousands of feet was a torment. The dapper clerks from Washington, sweating under the canvas, high-hatted the frontiersmen. Delays, technicalities and inefficiency made honest men furious, and failed to discourage the more daring, determined and unscrupulous "sooners."

Kansas, Texas and Arkansas furnished the greater numbers for the run, although every state in the Union was represented. The United States was in the depths of the worst depression it had ever known. That increased the hordes of adventurers. Thirty thousand persons registered at the booth below Arkansas City. Booths below the Kansas line at Caldwell, Hunnewell and Kiowa registered thirty-five thousand. On the south there were fifteen thousand at Orlando, a similar number at Hennessey, and seven thousand ready to start at Stillwater.

There are persons still living upon homesteads taken up in that rush who came from as far away as Wyoming with covered wagons, and saddle horses on which to make the actual run. Criminals fleeing from other states, indigents driven to desperation by the nation-wide depression, gamblers, land speculators and adventurous men and women of other types were sprinkled through the throngs of honest settlers. It was a day as mad as the day of the first rush of 1889, and with greater numbers starting upon a wider front.

As the hour of noon approached, hundreds of the fleetest and strongest saddle horses in half a dozen states, trained and conditioned for a race which might be half a mile or thirty miles across untilled prairie, were drawn up at the line in half a score of strategic points. Behind them were buckboards and covered wagons driven by old men, boys, wives and mothers who would follow the faster pace, to make homes upon the land which would be staked and claimed before nightfall.

Locomotive safety valves were popping off on every line which entered the Strip. A single Rock Island train of forty-two cattle cars carried a far greater weight of human freight than it had ever carried of beef. Other lines were equally crowded, with passenger cars jammed to the steps, and roofs sagging under the overflow. Speed of the trains was limited to give the horsemen a fair chance.

Pistol shots by troopers stationed in plain sight in front of the lines had been specified as the signal for the start. But at three different points premature shots from the excited throng started the stampede. At a crowded spot on the line south of Arkansas City the wild shot came four minutes before twelve. In the lead of the resulting break was one J. R. Hill, who had come all the way from New Jersey to take part in the rush. Vainly the soldiers tried to stop the movement. Two of them even pursued Hill for nearly a quarter of a mile, commanding him to stop. But the noise of the stampede shook the earth and air. Either Hill did not hear the commands or was too excited to heed. The troopers shot him dead. The stampede moved on. Similar incidents, minus the killing, occurred at the Hunnewell and Orlando fronts. They served only to prove the inadequacy of the military force.

In other areas the crowds lined up with great good nature while a trooper on horseback moved out upon the prairie in front of each wide front. Every eye was upon him. A puff of smoke from his revolver carried the signal to those beyond earshot. The long line moved almost with the unison of a field of thoroughbreds breaking from the barrier at Saratoga, Louisville or Santa Anita. And like the field of thoroughbreds it altered position quickly as it moved.

The faster horses gained the lead. The better stamina won its place. Wagons and buckboards and railroad trains made headway through the choking clouds of dust. Running horses stumbled in prairie-dog holes, and sometimes broke their legs. Wagons lost a wheel or a kingbolt at some rut or ravine. Some passengers were toppled from the steps and roofs of the overcrowded cars. But the waves of humanity moved on, engulfing the prairie, stopping where they must or where they would, and declaring the land upon which they stopped their own.

Halfway across the Strip the leaders met the winners of the race from the opposite direction. Before the sun set five million acres which had never known a plow, which had never known a white man except a traveling cowboy, a scouting outlaw, a buffalo hunter or a plodding trooper, had become a land of homesteads. No area of land so great had ever been settled with such speed and completeness in the history of the United States.

Charles F. Colcord was literally a leading figure in the Cherokee Strip stampede. Profiting by his experience in the rush which had founded Oklahoma City four years earlier, Colcord equipped himself with the best horse he could obtain. He selected the townsite of Perry as his goal, and headed toward it at full speed. He outdistanced every other man save one. As these two straightened for the final sprint, the big bay horse of his rival struck Colcord's horse on one flank and knocked it to its knees.

At the same instant the riders recognized each other. The man on the bay was George Parker, sheriff of Lincoln county. He pulled his horse to its haunches and sprang to Colcord's side. Colcord was already scrambling to his feet, unhurt. Parker, familiar with the survey of the townsite which he had visited in his official capacity, promptly made amends for the accident. "You're not hurt?" he asked. "I'm mighty glad of that. You stay right on this spot. This is the only fifty-foot business lot in town, and it's a corner besides."

Colcord stayed, and filed his claim. A day or so later

he bought a tract from its claimant one and one-half miles northwest of Perry for fifty dollars. Such deals were common in the rushes. Numerous persons made the runs and filed with the sole intention of selling their prior rights. Colcord established title to the acreage, built a modern eight-room house, and planted the first orchard in the Strip. He lived there with his family for several years, but returned eventually to Oklahoma City with such great profit to himself and his city as will be narrated in its place.

While similar incidents were occurring in a hundred places within the Strip at the same hour, the race was not always to the swift, or even to the lucky. Neither Secretary Hoke Smith's registration plan nor the military control entirely eliminated the "sooners." Lieutenant Colonel L. D. Parker was assigned the task of patrolling the border to hold back the rush until the zero hour, and also of covering the nine thousand square miles of territory to eject "sooners." For that colossal task he was allowed only eight troops of cavalry and four companies of infantry. It can hardly be wondered that some "sooners" escaped the troops. Illuminating examples have been recorded by E. G. Barnard, who was in the rush, and who wrote of it many years later in memoirs entitled *A Rider of the Cherokee Strip*, published by the Houghton-Mifflin Company.

Barnard, with a group of cowboy friends who had punched cattle over almost every section of the Strip, headed for an area of rich bottomland and woodland in Big Turkey Creek, near what is now Dover. It was some ten miles from the starting line. The cowboys had swift horses, accustomed to the terrain. They outdistanced all rivals. Yet before Barnard had been five minutes on the claim which he had selected an old man on a big grey mule which obviously had hardly been ridden at all appeared on the same site and started to drive his stakes. Only the menacing guns and unanimous word of the hard-boiled cowhands convinced him that his obvious status as a "sooner" would be proved to the authorities. Then he moved on. Others in similar situation, according to Barnard's observations, actually lathered their horses and mules with soap to indicate that they had made a furious run and were legitimately upon the land which they claimed.

They did not all meet with cowmen of the strength and assurance of Barnard and his friends. Doubtless many established their titles. Doubtless many honest settlers who had made the run and legitimately staked their homesteads were bluffed out by stronger and less scrupulous persons.

Trickery, bribery, and falsification of records were charged in the subsequent activities at the land offices, but most of the charges were disproved or simply allowed to drop. With only four land offices within the entire Strip, the confusion seemed hopeless. In an effort to speed the work a plan was adopted whereby the name of each applicant was recorded and he was given a serial number. That arrangement was expected to establish priority, and permit more time for the harassed clerks to make out the necessary papers in detail.

But there was a joker even in that. The lines before the numbering offices extended for hundreds of yards. If A had staked a certain claim at 2:10 P. M., but B filed at the land office upon the same claim at 2:09 P. M., the title must go to B. Still it had been necessary for the settlers to locate their claims before they could describe them for filing. So the first rush was to the land and the second rush to the land offices. In many cases more than one person had selected the same area. The first of those then to file at the land office would be the winner. The number issued to him as the line moved up would determine his priority or lack of priority. Places in line therefore were at a premium. Boys and women who had no claims to file established themselves in line and sold their places to the highest bidder until that practice was stopped by the weight of protest from those at the rear.

The system varied at the different land offices, as there were no official regulations governing that phase of the business. At one or two stations the lines organized themselves, dividing into platoons of twenty or twenty-five, electing a captain, and giving each platoon its proper place in the line and each individual his proper place in the platoon. With that arrangement individuals could leave the line for food or business, and be admitted to their proper places on return.

The numbering line at the Enid land office found itself moving with exasperating slowness. Suspicion grew quickly. Leaders among the men in waiting were elected to investigate. They reported promptly that "sooners" of a different type were finding their way into the rear of the numbering tent and bribing clerks to give them numbers in advance of the legitimate line. A committee of action followed the committee of investigation, met a clandestine group just entering the rear of the tent and threw them out bodily. That the clerks were merely admonished, and not beaten or hanged, may be considered evidence of the high good nature of the homesteaders.

Before the first week was past they were to meet a test of their stamina almost as severe as the stampede itself had been. Nearly all the millions of acres were covered with high dry grass. On the first night of settlement scores of campfires could be seen from any spot in the Strip. In the more thickly settled areas of the townsites the dry grass had been trampled flat in the first few hours, but on the open range it still swayed high in the breeze.

Some experienced homesteaders whose wagons and equipment had caught up with them promptly plowed lines around their claims, partly to define the area and partly as a break against the danger of fire which they had known in other prairie states. It was well that they did. Within another day or two the prairie was ablaze. Many careless homesteaders lost their entire equipment. Some were even forced to abandon their claims permanently because the fodder on which they had depended to feed their livestock through the winter had been destroyed.

Still the majority survived. The famous Strip was settled. Houses of lumber and logs gradually replaced the tents. Counties were organized under the government of Oklahoma Territory. Towns incorporated. Schools and courts were established. The mass drama of the last frontier for a time lapsed into individual dramas of human achievement against terrific odds.

CHAPTER VIII

THE WORST OF THE BAD MEN

OKLAHOMA TERRITORY, with Guthrie as its capital, comprised approximately half of what is now the state of Oklahoma. The eastern and southern area was still Indian country under Federal jurisdiction, without self-government except that of the restricted tribal councils. Perhaps it should be repeated that there never was an Indian territory with territorial government.

Although farms and cattle ranches were being improved, towns growing into small centers of civilization with courts, schools and churches, and several railroads criss crossing the area, Oklahoma Territory was still rude, rough country. The Indian territory, with the whites still barred from settlement, was a little more primitive.

The Centennial Exposition had been celebrated in the year of the Custer massacre. The Columbian Exposition, still known to fame as the World's Fair, was celebrated in Chicago in the year of the great rush into the Cherokee Strip. Those two expositions were a measure of the advance of civilization within the United States in general.

New York, Chicago and San Francisco were enjoying grand opera while the throb of the drums, the ritual chants of the aborigines and the songs of the lone cowboy were virtually the only man-made music within some forty-five million acres of the last frontier. Charles Duryea had built America's first gasoline-driven automobile. Electric lights had proved their practical value from the Atlantic to the Pacific. The telegraph had long since displaced the Pony Express across two thousand miles of wilderness. Railroad

87

trains were covering in four days a distance which had required four months of plodding hardship for covered wagons a few years earlier. The United States was settled, safe and content within its present boundaries. Only Oklahoma and the adjacent Indian territory still represented the old frontier.

To be sure, there were regions made more savage by Nature, such as the heights of the Rockies and the arid wastes of the western deserts, still unsettled, but there was left no other great area upon which human beings in large numbers might hope to make their homes and rear their children. And even there, though home fires had displaced council fires on a wide range, there were still vast and thinly populated areas of rugged hills and virgin lands.

The settlements had brought a rude and hardy people into the land. The railroads had brought not only the plows and harrows of civilization but the scourings of the criminal class of a score of self-respecting states. There were literally hundreds of hidden ranches where the criminals could find refuge between their forays.

Interesting pictures of how killers, cattle thieves, bank and railroad bandits, and their ilk lived and operated have been set down in numerous records. One of the most illuminating, I believe, may be found in E. G. Barnard's *A Rider of the Cherokee Strip*. Barnard entered the region as a boy in his teens. He was in it and of it through all the years from the great cattle drives, through the land rushes and homesteading era, to the present. There are in Oklahoma today numerous old men of similar experience, but Barnard's straightforward recital of his experiences as a youth in a frontier den of thieves is most illuminating. It indicates by its very naïveté how casually the pioneers accepted the desperadoes as long as their own property was not involved.

That, of course, has been the characteristic of all fron-

The first bank of Oklahoma City, marking the beginning of the end of America's last frontier.
(Courtesy *The Daily Oklahoman*)

tiers. In the San Francisco of the gold-rush period the bad men were tolerated or ignored by the more honestly busy citizens until all life and property were jeopardized. Then the vigilantes hanged the leaders and frightened away the others. The same course of events was recorded in the viciously tough frontier of Alder Gulch, Montana, and to a lesser extent in other western areas. It was a characteristic so much a part of frontier life that a few more examples may be permitted here to complete the picture of the last frontier.

At one stage of his checkered career as a cowboy, cook, and homesteader, young Barnard found a job with an outfit which he identifies as Cornell and Butler, supplying beef to the garrison at Fort Reno under government contract. At the time there was a gang of outlaws in the neighborhood making their headquarters at various cow-camps between the expeditions in which they drove stolen herds of Indian ponies to a market in Kansas. Few ranches refused to take them in, for fear that they would burn the grass and otherwise enforce a penalty. When President Cleveland ordered the cattlemen to evacuate all Indian lands, Cornell and Butler obtained a dispensation to remain until their contract for government beef expired. As the other cattlemen withdrew, the rustlers found themselves deprived of numerous places of retreat. More and more of them centered at the Cornell and Butler camp.

Barnard as camp cook had an unequaled opportunity to observe the outlaws. Night after night he watched them gamble with cowboys almost equally reckless. Six-shooters were always ready for any eventuality. On one occasion he overheard two of the most vicious men, Overby and Jackson, plot to trim a third desperado named Hunt. Overby and Jackson were professional gamblers wanted in Texas for murder. When Hunt was several hundred dollars ahead of the game Overby picked a quarrel with him and shot him through the neck.

The wounded man was taken to the government hospital at Fort Reno under an assumed name, and no complaint was made. Two months later, to the amazement of all concerned, he walked into the kitchen dugout. Overby and several others were at the table. Hunt was unarmed. When he attempted to borrow a gun he was refused. The man whom he asked agreed that Hunt ought to kill Overby, but not in that camp. There had been so much trouble traced to the camp and its outlaw residents that the group feared a raid by U. S. marshals if there was another shooting. That night most of the men in the bunkhouse went to sleep with their six-shooters in their hands. Soon afterward Overby disappeared.

Young Barnard had picked up so many details of crime on the part of various visitors that on several occasions he was near death at their hands because of their fear that he would inform the authorities. Still he stayed on, with fatalistic philosophy, until the Cornell and Butler contract with the government ran out and their remaining herd was removed. The young man's attitude was a common one of the time. The territory was a fertile field for crime, but too many criminals were tilling it and the harvests generally were small.

A similar situation existed in the Indian territory. A hundred killers and bandits have left their names in the archives of that region, chiefly in the court records at Fort Smith. Most notorious, brutal and desperate of the lot was one Crawford Goldsby, who has come down in the history of the frontier as "Cherokee Bill." He was the son of a soldier stationed at Fort Concho, Texas, and a mother who was half negro, one-fourth Cherokee and one-fourth white. From his father he also had a strain of the fierce Sioux blood. Either that mixture or his environment, or both, produced a vicious human being.

Goldsby enjoyed some advantages of education. He

had three years of schooling at Cherokee, Kansas, and two years in Carlisle, Pennsylvania. But probably more important in the formation of his character was his mixed-blood mother's oft-repeated counsel: "Stand up for your rights; don't let anybody impose on you." That, doubtless, is excellent advice for a well-bred child who has been taught the difference between rights and license, but for one handicapped by inferior breeding and environment it is likely to produce a chip-on-the-shoulder attitude, extremely dangerous.

The boy returned from the Carlisle school to Fort Gibson to find his mother remarried, with little interest in his activities. He ran loose with the worst boys in the community. He developed a taste for liquor. By the time he was eighteen years old he was a big brute of a youngster who had never been curbed, and who resented any attempt to curb him. When he quarreled with one Jake Lewis, a negro nearly twice his age, at a dance at Fort Gibson, he received a terrific beating. From that time onward he never trusted to his fists. Two days later he met Lewis and shot him through the body. When Lewis ran the boy shot him again. It appeared to be a finished job of premeditated murder. Goldsby fled. In a region filled with bad men, his position was established.

That was in the spring of 1894. Two months later, when a Federal payment of $265.70 to each member of the Cherokee Nation in settlement of more than six million dollars due for the purchase of the Cherokee Strip was started at Tahlequah, Goldsby, with Jim and Bill Cook whose Cherokee blood also gave them rights in the payment, decided to collect.

Goldsby knew that he was wanted for the shooting of Lewis. One of the Cooks was wanted on a larceny charge. They knew they might be arrested in Tahlequah, so they stopped at a place known as the Halfway House, about fourteen miles from Tahlequah on the Fort Gibson road. It was kept by one Effie Crittenden, estranged wife of Dick Crittenden, also a mixed-blood Cherokee.

The three desperadoes induced Effie Crittenden to take their orders for the cash allotment into Tahlequah and collect the money. With the cash in hand they remained at the Halfway House to spend some of it with their hostess for liquor, planning to slip away within a day or two. Just at nightfall a posse including Dick and Zeke Crittenden, husband and brother-in-law of the roadhouse landlady, and one Sequoyah Houston, galloped noisily up toward the doorway, and opened fire upon Goldsby who was seated outside. Some members of the posse were as drunk as the men whom they sought to arrest. Virtue on the frontier was not always entirely on the side of the law. Goldsby, unwounded, rifle in hand, dodged behind a corner of the building and returned the fire. The Cook brothers joined the battle. Jim Cook was wounded seven times. Sequoyah Houston was shot dead. As Houston dropped, the sheriff and four of his posse took to their horses' heels, leaving the Crittendens to continue the battle, if they wished.

Apparently the Crittendens were bold men. It has been reported that they planned the raid in the hope that Dick's estranged wife, Effie, would be summarily eliminated. In any event they continued shooting from such cover as they could find until the desperadoes, including the wounded Jim Cook, escaped in the darkness.

It was in that battle that young Goldsby gained the alias which clung to him for the remainder of his brief life. On the day after the fight a woman in the house was asked if Crawford Goldsby had been there. She answered, "No, it was Cherokee Bill."

Both the Cook brothers and Cherokee Bill were then branded as outlaws " on the scout." The notorious Cook gang, including Cherokee Bill and half a dozen others, of

whom Sam McWilliams, "the Verdigris Kid," and Jim
French became the most widely known, began the robbery
and terrorization of the countryside.

Early in that same autumn two men, later identified as
Cherokee Bill and either Jim French or the Verdigris Kid,
galloped into the little town of Lenapah, a station on what
is now the Missouri Pacific railroad, about half way between Coffeyville and Tulsa. The pair dismounted in
front of the general store kept by Schufelt & Son. Their
general appearance was that of scores of cowboys who
came frequently to the store to trade. No one paid any
special attention to them until they leveled rifles upon the
men in the store and commanded, "Hands up!"

Cherokee Bill entered the store while his companion
held the horses at the door and occasionally fired a shot
along the street to keep the citizens within doors. Bill
marched John Schufelt to the safe at the rear, forced him
to open it, and scooped the money into his pockets. Then
he forced Schufelt to make a bundle of various items
which took his eye, and made his way to the street. There
his companion suggested that they needed more cartridges,
and Bill re-entered the store to obtain them. As he moved
to the rear he happened to glance through a side window,
and saw a man named Ernest Melton peering from a window in a small restaurant across a vacant lot. As calmly
as he might have killed a fly, Cherokee Bill shot Melton
through the head. Then he picked up the needed cartridges, rejoined his companion and galloped away in a
cloud of dust.

In a period of less than three months Cherokee Bill par-
ticipated in the robbery of Scales' store at Wetumpka,
Creek Nation; robbery of a train at Red Fork, Creek Nation; robbery of Parkinson's store, Okmulgee, robbery of
an express office at Choteau, robbery of a train at Coretta,
and various other robberies, in some of which murder
was an incident. He was a bad man in bad country.

His capture was brought about as the capture of so
many of his criminal successors of a later day has been
brought about—through his weakness for women. Deputy
U. S. Marshal Smith, operating out of Fort Smith, discovered that Cherokee Bill had developed a passion for
one Maggie Glass, a mixed-blood who lived near the home
of Isaac Rogers, an erstwhile deputy marshal, also of
mixed blood. Deputy Smith arranged with Rogers to in-
vite Cherokee Bill to meet the girl at his home.

Bill and the girl kept the appointment, though both had
become somewhat suspicious of Rogers' good faith. When
the girl warned her lover Cherokee Bill merely tightened
his guard, kept his gun within easy reach, and told her
that he would kill his host at the first treacherous move.
The outlaw even refused the whiskey which Rogers
pressed upon him. No better evidence of his suspicion
could be adduced.

Most of the night passed in a card game in which Clint
Scales, a neighbor who had joined the conspiracy for
Cherokee Bill's capture, was a participant. The outlaw
sat with his back to a wall and his gun at hand. It must
have been a strained party. It was four o'clock in the
morning before the game broke up and the participants
went to bed with Cherokee Bill and Rogers stretched on
the same mattress. Breakfast did not greatly relieve the
situation.

Rogers' instructions were to take the outlaw alive, but
he was beginning to doubt that he could take him either
alive or dead. The three men sat together before the open
fireplace after the meal. Cherokee Bill finally suggested
that it was time for him to leave. Rogers urged him to
stay, and supported the invitation with a promise of a
chicken dinner. That promise he intended to serve a
double purpose, half of which was to get the girl, Maggie
Glass, out of the house. He gave her a dollar to buy chickens from a neighbor half a mile away, and she departed.

A few minutes later Cherokee Bill bent to the hearth for an ember with which to light a cigarette. It was the first move in more than twelve hours in which he had relaxed his watchfulness. Rogers seized the opportunity and struck the desperado across the back of the head with a stick of firewood. Telling of the incident later he declared he hit hard enough to kill an ordinary man, but merely succeeded in knocking Cherokee Bill to his knees.

Rogers and Scales together then jumped upon the outlaw while Rogers' wife seized the rifle, knocked aside at the first blow. The battle on the floor lasted fully twenty minutes, according to Rogers' account, before the two men managed to get handcuffs on the desperado. They loaded him into a wagon and started to Nowata to turn him over to the authorities. On the way Cherokee Bill again revealed his tremendous strength by actually breaking his handcuffs, but Rogers held him in place with a gun, and he was handed over to Deputies Smith and Lawson. With an additional guard, including Dick and Zeke Crittenden, he was promptly removed to jail at Fort Smith to await trial for the murder of Ernest Melton.

Another similarity between the crime and justice of the frontier of forty years ago and the crime and justice of the city ganglands of prohibition days was demonstrated at the trial. J. Warren Reed, a famous and successful criminal lawyer of Fort Smith, and almost a prototype of such later defenders of criminals as Earl Rogers of California and William J. Fallon of New York, took charge of the defense. He claimed an alibi, attempting to prove that at the time of the murder Cherokee Bill was in Fort Gibson, ninety miles from the scene of the crime. But when numerous witnesses identified the prisoner as the man whom they had seen do the shooting, the jury ignored the heart-throbbing appeal of counsel for "the poor Indian boy." Even the weeping mother who had been seated in the courtroom for emotional effect upon the jurors was ineffectual. The verdict was, "Guilty, as charged." The defendant's mother and sister broke into wails of grief.

"What's the matter with you? I'm not a dead man yet, by a long ways," Cherokee Bill harshly admonished his mother. He spoke some truth.

Reed immediately demanded a new trial. The application was denied. Judge Parker was accustomed to seeing the men who were convicted in his court pay the penalty on the gallows. But Reed had saved the life of more than one vicious criminal. When a new trial for Cherokee Bill was denied, Reed appealed to the Supreme Court of the United States. When that appeal was denied he turned to President Cleveland for a pardon.

The delays caused by those appeals gave Cherokee Bill the opportunity for his final atrocious murder, coupled with the wildest outbreak ever to occur in the Fort Smith jail. At the time, June, 1895, there were fifty-nine men under sentence of death in the jail. In addition to that list of murderers, indicative of conditions upon the frontier, there were innumerable train robbers, bank robbers, highwaymen and other criminals, awaiting trial or removal to penitentiaries. And in the face of that situation a thrifty government had reduced the number of guards to a minimum. A fully loaded revolver was found in Cherokee Bill's cell, and nine extra cartridges in the cell above, but no extra guards were placed, no special precautions taken.

It was the routine practice of the jail to send all prisoners from the liberty of the corridors on each of the three floors into their cells to be locked up at 6:15 P. M., immediately after the night guards came on duty. The first movement of the lock-up was the pulling of a lever which threw a bolt at the top of each cell door in the row. One turnkey then put aside his weapons and entered the inner corridor to lock each cell individually with a key while a second guard, fully armed, walked down the parallel corridor outside the bars to protect him in any emergency.

There was one grave error in this arrangement which Cherokee Bill or one of his fellow murderers had had the wit to see and the daring to use in planning a concerted jail-break.

The lever-operated bolt, or "brake," as it was called, could be reopened by a man with a stick in the cell at the north end of the tier before the turnkey moved along to lock each grated door. On the night of the tragedy the "brake" had been thrown back. Although the guards were not aware of it, every cell on the corridor on which Cherokee Bill was held needed only a push from the desperate man within to open it. In this situation Turnkey Eoff moved along the cell row throwing each individual bolt while Guard Lawrence Keating moved along the adjoining corridor for his protection. Meanwhile Cherokee Bill with a six-gun in his hand waited behind his unlocked door.

When Turnkey Eoff reached the cell next to Cherokee Bill's he had trouble in inserting his key. A wad of paper had been pushed into the keyhole. Eoff remarked to Keating that there was something wrong. Keating moved closer. At that instant Cherokee Bill leaped from his cell, thrust his gun through the outside row of bars at Keating and shouted: "Throw 'em up, and give me your gun!"

Had Keating complied Cherokee Bill would have had both guards unarmed, at his mercy while he unlocked the door out of the corridor and freed himself and his fellow prisoners. But Keating, a brave and competent man with nine years of experience in the jail, raised his gun to shoot. Cherokee Bill promptly drilled him with two bullets. Keating staggered backward. Eoff, within a stride of the murderer, was unarmed and helpless. He turned and ran, dodging around the end of the corridor as Cherokee Bill sent two more bullets after him. At the same time George Pearce, another murderer under sentence of death, leaped from his cell, armed with a club. He rushed after

Eoff evidently expecting to seize the keys and open the outside door of the corridor to freedom. But Eoff had left the keys stuck in the lock where he had first met difficulty.

Before Pearce could make that discovery the alarm of battle had brought two more guards on the run. As Keating collapsed, dying, Guard McConnell seized his gun and fired twice through the bars at Cherokee Bill. The murderer returned the fire, dodged around the end of the corridor and fired at guards approaching from that side. Then he darted into his cell, reloaded his gun and came out shooting. Deputy Marshal Bruner and half a dozen other men came at a run from other parts of the jail, seized any arms available and raked the corridor with bullets and buckshot. The powder smoke was so dense that no one could define a target. Prisoners on the upper tiers were screaming and beating upon their bars in a frenzy. Hell was popping.

At last one Henry Starr, a prisoner on the murderers' tier, volunteered to go to Cherokee Bill's cell and get his revolver if the guards would cease firing. He did so after a brief argument to convince the murderer that the last hope of a successful jail delivery had vanished.

Hundreds of persons from the town, attracted by the fusillade and screaming uproar, had gathered outside the jail. When Keating, the slain guard, was carried out, the crowd muttered vengeance. He had been a popular man in Fort Smith, with a wife and four children. Even the armed guards and officers within the jail for a moment considered shooting down the murderer. But cooler heads managed to prevent a lynching.

Although Cherokee Bill's appeal to President Cleveland for pardon still remained to be denied, the murderer was again promptly placed on trial before Judge Parker. The jury found him guilty after only thirteen minutes of deliberation. Again he was sentenced to die upon the gallows. But again his counsel appealed to the Supreme Court, and a second stay of execution had to be issued. It was a disappointment to more than five hundred persons who had assembled from a radius of seventy-five miles to see him hanged. Hangings were public functions in those days. Cherokee Bill had become the most notorious figure in all the Indian territory under the jurisdiction of Judge Parker's court. He was generally believed to have killed thirteen men, although he was convicted of only two murders.

When denials of all the appeals had been entered and the day of execution finally arrived, March 17, 1896, the town of Fort Smith was crowded with thousands who hoped to witness the hanging. The jail yard where the gallows stood could not begin to accommodate the crowd. One hundred tickets were issued, but several hundred persons without tickets climbed the walls and dropped into the throng.

In his last moments Cherokee Bill, drunkard, wastrel, robber and murderer that he was, displayed far more dignity than many of those gathered to see him die. He marched with firm stride between guards across the jail yard and up the steps of the gallows. When one of the guards asked if he wanted to say anything he replied, "No, I came here to die, not to make a speech." When the marshal after reading the death warrant asked if he had anything to say he answered in a low tone, "No; except that—I wish the priest would pray."

The priest complied briefly. Cherokee Bill stepped upon the trap. Deputies bound his arms and legs, adjusted the noose, pulled the black cap over his face. Guard Eoff, who had figured so prominently in the attempted jail break, released the trap. Cherokee Bill's career came to an abrupt and logical end. He was just past twenty years old. He had crowded more crime into two years than any other man in the turbulent history of the last frontier crowded into a lifetime.

CHAPTER IX

DRAMA OF SETTLEMENT

OKLAHOMA and the Indian territory were beginning to settle into some of the more conventional ways of civilization after the hanging of Cherokee Bill. They still had a long way to go. Oklahoma City was a cow-town sprawled upon the prairie. Tulsa was a railroad junction, scarring the landscape beside the Arkansas River. Guthrie was the capital of Oklahoma Territory. Muskogee, in effect, was the capital of the Five Civilized Tribes. Pawhuska was the chief village of the independent Osage Nation. Various other groups of Indians, Sac and Fox, Iowa and Kickapoo, Wichita and Caddo, Comanche and Kiowa, still maintained a semblance of self-government under Federal control within Oklahoma Territory.

The situation was peculiar. Authority was divided among three broad groups, the territorial government of Oklahoma, the Indians and the Federal government which still theoretically protected the Indians and maintained control of large areas.

The first years of development of Old Oklahoma after the rush of 1889 were years of struggle and disappoint- ment. The first crops were meager. The cattlemen who were still leasing grass lands from Indians or grazing their herds on lands not yet opened to settlement resented the coming of the "nesters," as they called the farmers who were trying to establish homesteads. Numerous and fre- quently violent were the clashes between cattlemen and "nesters." Cattlemen who had made fortunes through cheap grazing privileges had acquired the belief that they

100

had some God-given right to the range. The movement
of a plow across a section of such land was an outrage, to
be resented and punished by violence in areas where territorial courts were not entirely effective.

Some of those same cattle kings had encroached upon
the original Indian lands in similar manner but on a far
wider scale than the "nesters" were encroaching. They
had justified themselves with the specious argument that
the Indians had no use for the lands after the buffalo herds
had been destroyed. It was as good an argument as any,
but when the incoming settlers developed a similar theory
against the cattlemen it prompted reprisals.

Herds stampeded across the isolated farms, destroying
growing crops. Tough ranch managers intimidated "nesters" coming into the scattered settlements to buy their
meager supplies. Cowboys doubling as gamblers enticed
the more reckless farmers into poker and monte games,
and occasionally cleaned them of their last cent and even
their horses and wagons, setting them afoot with no
chance to go on with the work of proving-up on their
homestead claims. Politics came in with the election of
sheriffs and the appointment of marshals who would give
the cattlemen the best of any deal. The result was a situation which has formed the basic plot for thousands of fictional tales of the frontier, still popular.

Quite recently I have read one such story in a magazine
which boasts a circulation in the millions. A stirring tale.
The girl comes from her conventional home in the East
to marry her childhood sweetheart in the West. She learns
that he has just died with his boots on. The bronzed hero
who tells her is apparently about to do the same if he continues to defy the leading cattleman's ultimatum to abandon his homestead and quit the country the next day. His
cabin and barn have already been burned. He is a cowboy who has broken the sacred tradition of the range by
becoming a "nester." Three competent gunmen from the

cattle king's forces are in town to see that orders are obeyed. All the other citizens are standing back, ready to dodge a bullet. The bronzed hero has not cared very much, except that he resents being forced to take orders, until the girl's quiet strength inspires him with new ambition. He cannot run away while her clear blue eyes are upon him. He declines to board the last train out. Bullets begin to fly. The hero kills the chief gunman, wounds a second and puts the third to flight. Clinch and fadeout.

That is more or less the way things actually happened on the last frontier, as they had happened in earlier days upon the plains and in distant California. It was not always quite so dramatic. Frequently it was merely a heartbreaking story of toil and hardship and poverty. By the time the Cherokee Strip was opened the entire United States was suffering in the financial panic of 1893, beginning four years of depression with hardship as devastating as the recent nation-wide depression. In the nineties men attempted to stand upon their own feet, not to lean upon a benevolent paternalistic government. It was a heartbreaking experience, but it built character and an invaluable pride in self-sufficiency. The homesteaders of Oklahoma used their ingenuity. Mostly they were honest. Sometimes they were tricky. Occasionally they resorted to fraud. But they got along without "relief" and without charity.

Charity would have been impossible in a neighborhood where all the villagers and farmers assembled in a sod schoolhouse for church services could show but sixty-five cents among them.

Thousands sustained themselves largely on cow peas, turnips, kaffir corn ground in coffee mills and an occasional wild turkey, pigeon, rabbit or deer from the swiftly vanishing supply of game. Proper clothing, farming implements, even seed for planting, were almost impossible to obtain. Two railroads, the Santa Fe and Rock Island, with a practical eye to future freight, furnished seed wheat to some farmers, and eventually profited.

Watermelons and turnips were the most bountiful crops in the summer of 1894, the first summer of attempted farming in the Cherokee Strip. There was no market. Millions of melons rotted in the fields. Turnips were down to a few cents a bushel, and no demand at that. But there were ingenious men among the homesteaders. George Rainey has revealed some illuminating incidents in a volume entitled *The Cherokee Strip,* issued by the Co-Operative Publishing Company, Guthrie, Oklahoma, in 1933.

At the lowest point of the market one day a boy slouched into a grocery store owned by H. C. Kennedy in Enid and purchased ten cents' worth of crackers and cheese. He seated himself on a woodbox behind the stove and munched hungrily. A roughly clad man, evidently a farmer, entered the store and asked the proprietor if he had any turnips to sell, explaining that he had a lot of hogs at home and that he had been told boiled turnips made excellent hog feed.

He wanted a wagonload, and said he would pay twenty-five cents a bushel. The storekeeper knew that he could buy turnips for ten cents a bushel if he could find them. That seemed unlikely as the local stores had long since ceased to stock an article for which there was no call. But he visualized a profit of one hundred and fifty per cent, and was willing to try. He told the farmer to come back in an hour. The man shambled out.

The boy, stuffing the last of the crackers and cheese into his mouth, came out from behind the stove. "Say mister," said he, "when I was comin' up the street I seen a whole wagonload of turnips down near the depot. Them crackers an' cheese was good." And he ambled out into the wind-blown street.

Kennedy locked his door and went on his search. Around a corner, sure enough, he found a man wrapped in a ragged overcoat, crouched upon the seat of a wagon loaded with turnips. "Want to sell your turnips?" the storekeeper asked.

"That's what I brung 'em in fur," said the man, "but I ain't had no luck."

"I can use 'em," the storekeeper answered. "Ten cents a bushel."

"No, I'd ruther haul 'em home an' feed 'em to the hogs. Fifteen cents is the least would pay me."

The deal was made at fifteen cents. The turnips were unloaded at the grocery store. The farmer collected his money and drove slowly away. Kennedy waited for the wholesale customer to appear and take the turnips off his hands at twenty-five cents a bushel. He waited in vain. Neither the customer, the driver nor the boy was seen again in Enid.

In another grocery store in a small settlement a man approached the proprietor with a hard-luck story. "We're all out of grub at my place," he said. "The ol' woman an' the younguns is hongry. I ain't got a damn' red cent, mister, but they's a fat shoat out in front in my wagon that I'd like to trade in for some groceries."

The storekeeper examined the pig, named his own price, made up the payment in groceries and told the farmer where to unload the shoat. When the grocer went at nightfall to feed the animal it was not in the pen. Some investigation disclosed that it was not in the pens of two other grocerymen in the town who had been worked for a bill of groceries in similar manner.

Settlers who did not possess the originality, ingenuity and unscrupulousness of the turnip salesman or the pig salesman did the best they could. Some made a precarious living by gathering bones of buffalo or cattle, long since slaughtered, starved or frozen on the prairie, and selling them to fertilizer factories in Kansas City, St. Louis and

Chicago. Some who had established themselves within reach of blackjack or other timber cut firewood and sold it in the towns at sixty cents a wagonload. Perhaps three loads of wood could be cut and hauled to market in a week. That brought an average of thirty cents a day for the hardest kind of labor. But $1.80 worth of groceries to supplement the limited supplies in the root cellar and smokehouse could sustain life for a family on the last frontier.

The struggle was not only between man and Nature but between man and man, and town and town. No frontier area so vast as the Cherokee Strip, for example, could be occupied throughout its millions of acres in a single day without the clashing of groups as well as the clashing of individuals. The scenes of confusion have already been pictured. They were extended through the rivalry of various towns or groups.

The railroads which crossed the Strip from north to south were partly to blame for that rivalry and confusion. The incompetence of the Federal bureaucracy which had planned the settlement was equally to blame.

In preparation for the opening the Secretary of the Interior had divided the Strip into seven counties, and designated a site for each county seat. For some mysterious reason or lack of reason the county seat of what was originally "O" county, later Garfield county, was designated at a point three miles south of the Rock Island Railroad's established depot, warehouse, sidings and cattle pens, known as Enid. That shipping center had been in operation prior to the government's cancellation of the cattlemen's leases of grazing rights in Strip lands.

The railroad maintained that the designation of a townsite on open prairie three miles from an already established shipping center was absurd. Therefore on the day of the opening rush the Rock Island trains moving northward into the Strip refused to stop at the designated site. An idea of the immediate effect of that action may be gathered from a description of those land-rush trains.

Each Rock Island locomotive hauled forty-two cattle cars and a caboose. Every car was jammed like a New York subway train in the rush hour, with an extra load of hundreds packed on the roofs and hanging to the side ladders. Speed was limited to fifteen miles an hour to give the homesteaders on horseback something like an even break. The trains were to stop every five miles and give the passengers a chance to scatter and stake such claims as they might desire. A majority of the passengers had their eyes on townsites, while a majority of the cross-country stampeders sought farm lands.

Northbound passengers from Hennessey who sought lots in the designated county seat below Enid were carried three miles past their goal, and forced to walk back. The excitement and turmoil can be imagined. Many passengers leaped from the tops of the cars. Those packed like beeves within struggled to escape. Some, including one woman, Mrs. E. W. Van Brunt, leaped from doorways and staked their lots. Only two persons were injured, one breaking a leg and one an ankle. The bulk of the passengers were carried on to the railroad siding, and there forced to choose between a three-mile hike back to the designated townsite or a lot where they stood, with possible advantages of railroad service.

The immediate result was two towns within three miles of each other. The south town had a postoffice but no depot. The north town had a depot, freight shed and siding, but no postoffice. The south town called the other North Enid, although it had been the original Enid, so named when the railroad built the first cattle pens there. The north town called its rival South Enid. Freight for the south town had to be hauled in wagons from the north town, and residents of the north town had to watch their incoming mail taken from under their eyes to the post-office three miles away where they must follow to claim it.

Each group jeered and derided the others when they came, as they were forced to come almost daily, to collect their mail or haul their freight.

At the same time similar conditions disturbed the peaceful improvement of the rival communities which marked the railroad station of Round Pond and the designated county seat of Pondcreek, to the north. Residents of Pondcreek resorted to desperate efforts to force the trains to stop at their village. On one occasion a group of house-movers stalled a small frame house on the track. The Rock Island locomotive driver had been well advised of his duty. He pulled the throttle wide open, ducked his head within the cab and went through the house as though it were a movie set, then unknown to the world. But the Pond-creekers were not to be denied.

Dynamite blasts wrecking two small bridges across the Salt Fork River brought residents of the town forth at a run on a June morning of 1894. Mr. Rainey reports that a member of the running crowd checked its speed by suggesting that they should give the fellow a chance to get away. When a man was seen running from the scene of the wreckage no one followed. Shortly afterward a railroad bridge was burned near Kremlin. But the trains did not stop at Pondcreek. A few weeks later some two hundred men tore up a section of railroad track in front of an approaching trainload of cattle.

When one resident tried to flag the train with a red petticoat snatched from a clothesline the engine driver ignored him. That was an outworn trick in Pondcreek. The train sped into the trap and piled twelve cattle cars in a heap, killing most of the animals. Eighty residents of Pondcreek were arrested on Federal warrants for that incident, but none was punished.

The situation was becoming a bit too rough and dangerous for the railroad and the settlers alike. The railroad was losing business both in passengers and freight. Development of the county-seat towns was being restricted. Petitions were sent to Congress to compel the railroad company to build stations and stop trains at the towns in dispute. Oklahoma had only territorial representation in Congress, but it had friends. The House passed the proposed bill promptly. The Senate delayed.

But when a freight train was wrecked a mile south of Enid by the sawing of timbers on a bridge the Rock Island's chief of counsel advised the lobbyists in Washington to cease opposition to the proposed legislation. It was passed immediately. The south town of Enid, with the new benefit of railroad service added to its advantages as seat of the county government and U. S. land office, quickly absorbed most of the business, and even the buildings, of the north town. Round Pond likewise gave way to Pondcreek.

Another drama of frontier life had reached its climax and anticlimax. It is an entertaining memory to numerous original settlers still living in that section of Oklahoma. By 1897, with the final fading of the panic of '93, an excellent crop and fair prices came together. The worst hardships of the days of settlement were past.

Simultaneously another force was exerting itself. The Indian question had long since ceased to be a problem of general interest to the people of the United States, but it was still of vital importance to thousands of white and red men living in Oklahoma Territory and adjoining Indian territory. The railroads had obtained a large part of what they wanted. Land-hungry settlers had done likewise. But both railroads and settlers yearned for additional advantages and safeguards of statehood. The Indians were not so anxious to exchange the slight independence which they still enjoyed for the questionable restrictions and privileges of U. S. citizenship within a state. On two occasions they managed to block action aimed at placing their territory within a new state to be called Oklahoma.

In that situation President Cleveland, in 1893, appointed Senator Henry L. Dawes of Massachusetts, who had long taken a keen interest in Indian affairs, to head a commission with the purpose of persuading the Five Tribes to exchange their tribal lands for individual allotments. That arrangement would automatically destroy the tribes as nations.

The first task of the Dawes Commission was to convince the Five Tribes that their interests would be best advanced by abandoning their already restricted tribal governments in return for a new status as citizens of the United States. That, it was explained, would give them all the privileges and advantages of citizenship, including equality before the courts, votes at all elections, freedom of the press, speech and worship, and free schooling. In return for that the tribes should submit to cancellation of title to their lands and accept individual allotments of acreage, cash, or both.

The Indians had been hearing such specious arguments and promises for nearly a century. They had innumerable reasons, based upon tragic experience, to doubt the good faith of the Federal government. But the pressure upon them was terrific. As revealed in an earlier chapter, their own leaders were not always the "noble red men" of fiction or sentimentalized history. They were human, as the whites were human. Some were well-to-do and ambitious. Some were poverty-stricken and lazy. Many were vain and egotistical, common characteristics of all the Indians in all the country in all the centuries of their exploitation by the whites. Some were stupid. Some were criminal. Nearly every one had a weakness through which he could be approached.

The result was a final breakdown of tribal solidarity. The Choctaw and Chickasaw nations were first to agree, in 1897, to the proposition put by the Dawes Commission. The Creeks followed quickly. The following year the Curtis Act passed by Congress provided for abolition of tribal courts and preparation of a roll of Indian citizens with allotments to each under a survey and appraisal to be made by the Dawes Commission. The Cherokees and Seminoles were to come in soon. All the tribal governments were to be discontinued in eight years.

Some of the newly ceded lands were to be opened to settlement. Large areas, including mineral lands and timber lands, were reserved from allotment, eventually to be sold and the price divided between members of the tribes. The Federal government retained a large measure of control, especially over mineral rights, to be exercised through the Secretary of the Interior.

Although the task of the Dawes Commission was colossal, requiring the services of five hundred employes and twelve years of application before it was completed, the greatest obstacles to formation of a new state seemed to have been eliminated by the Curtis Act.

The principal chiefs of the Indians were politically ambitious. They knew that the officials of Oklahoma Territory were already in line for preference in a proposed new state. They knew they could not hope to exert any effective influence in opposition to the whites. But they could hope to direct and profit by the direction of an independent state occupied almost exclusively by their own people.

John F. Brown of the Seminoles, W. C. Rogers of the Cherokees, Green McCurtin of the Choctaws and Pleasant Porter of the Creeks, each the principal chief of his own nation, called a convention to meet at Muskogee for the purpose of adopting a constitution for their own proposed new state of Sequoyah. The constitution was submitted to their people, and approved, but by a light vote indicating little interest. It was soon rejected by Congress.

The last concerted effort of the Indians to maintain their integrity had failed. The red man was taking the white man's road.

In the meantime a tremendous new factor had slipped quietly, unobtrusively, but with potentialities beyond the wildest imagination, into the economic, social and political life of the Indian territory. Oil!

CHAPTER X

THEN CAME OIL

THE history of oil in Oklahoma is older than the state, older than the territory, older than the petroleum business in the United States. In the year 1853, six years before Edwin L. Drake opened the first commercial oil well on this continent near Titusville, Pennsylvania, a Federal Indian agent accredited to the Choctaw Nation reported to the government:

> "The oil springs in this nation are attracting considerable attention, as they are said to be a remedy for all chronic diseases. Rheumatism stands no chance at all, and the worst cases of dropsy yield to its effects. The fact is that it cures anything that has been tried. A great many Texans visit these springs, and some from Arkansas. They are situated at the foot of the Wichita mountains on Washita river."

The government did nothing about it. Coal and firewood, and some waterwheels, furnished the heat and manufacturing power of the nation. Whale-oil lamps, coal-oil lamps and candles furnished illumination in mansion and hovel alike. There was no reason why the government should be interested in a promised cure for rheumatism and dropsy. It had not developed the paternalistic sociological theories which were to blossom in the next three-quarters of a century.

Even when Drake brought in the first American oil well, in Pennsylvania, August 27, 1859, the government was not interested. Oil had value, but before the development of the gasoline engine and the attendant need for improved lubrication of high-speed machinery the value was limited as the market was limited. John D. Rockefeller's imagination and methods had not yet become a factor. The best that the discoverer of oil in Kansas could do, twenty years after the Indian agent's announcement of oil in the Choctaw country, was to skim it from the surface of pools on his farm where it collected, and sell it for axle grease.

But a decade of development of the Pennsylvania fields had revealed fortunes, real and potential, in the oil business. Pennsylvania oil men had moved westward through drilling operations in Ohio, Indiana and Illinois, sinking larger sums into the ground than they took out, but occasionally striking oil with profit to the lucky ones. Uses and understanding of oil were expanding. Possibilities of quick fortunes were appealing to more and more persons.

One Robert M. Darden came from Missouri into the Indian country and organized the first oil company in that area in 1872. It was incorporated as the Chickasaw Oil Company. Leading Chickasaws and Choctaws met at the home of Winchester Colbert in old Pontotoc County, Chickasaw Nation, and signed a contract with Darden to permit oil development on their lands in return for a royalty of one-half the total production, in barrels at the wells. The Indians were not too stupid. That was the largest royalty percentage any of them ever stipulated. That it was the least profitable was not their fault.

The M. K. & T. Railroad was building through the territory at the time. Coal, not oil, was the fuel demanded. The Darden enterprise produced no oil. In the next ten years, however, rich wells began to produce in Kansas. Markets in general were improving. Prospectors and promoters were extending their range. In 1883 Dr. H. W. Faucett of New York signed an agreement with the

Choctaw Nation under which he was privileged to drill for oil on a royalty basis.

Faucett started to drill on Boggy River, twelve miles west of Atoka in the Choctaw Nation, in 1885. He organized the National Oil Company with St. Louis capital. Drilling was expensive, partly because of poor tools and methods and partly because of high costs of transportation. Faucett was impeded by a constant struggle for more funds. Slight traces of oil and gas stimulated interest only intermittently. The Choctaw Oil and Refining Company displaced the original company. The well reached a depth of 1400 feet—a very deep well for those days—but it was a dry hole. Faucett died and the Choctaw Oil and Refining Company closed out.

Michael Cudahy of Omaha, founder of the great packing company which still bears his name, obtained a blanket lease on all the lands of the Creek Nation in 1884, and sunk a test well near Muskogee. Some signs of oil at 1,120 feet depth soon petered out. At 1,800 feet the well was abandoned as a duster.

It was in August, 1889, just thirty years after the opening of the first successful well in the United States, in Pennsylvania, that one Edward Byrd opened the first producing well in the Indian country. That was on a branch of Spencer Creek, west of Chelsea, then in the Cherokee Nation, now northeastern Rogers County. It was only thirty-six feet deep and produced only half a barrel of oil a day. Even with oil prospectors from Kansas and points east taking more and more interest in the region, it could hardly be described as a commercial well.

But Byrd held an oil lease on ninety-four thousand acres, signed by Chief Bushyhead of the Cherokees. He drilled a second well a little more than twice as deep and pumped three barrels a day. A third hole was lost, and a fourth, north of the creek, struck oil at ninety-six feet. It produced five barrels a day. It was that well, opened in

November, 1889, that students of oil history cite as the first commercial well in Oklahoma.

By that time the Interior Department was beginning to pay more attention to the Indians' rights in the minerals underlying their lands. The validity of Byrd's lease was questioned. Red tape entangled the drilling machinery. Byrd shortly abandoned the whole enterprise in disgust and moved to Texas.

When news reached George B. Keeler, operating a trading post on Caney River, that oil had been discovered at Neodesha, Kansas, he recalled that he and Jasper Exendine, as far back as 1875, had noted evidence of oil on Sand Creek when their horses refused to drink. At that time there had been no interest in oil and no drilling outfits available. Keeler had forgotten the incident until the strike at Neodesha revived his interest with a rush.

Keeler and two other Bartlesville pioneers, William Johnstone and F. M. Overlees, hurried to Tahlequah, headquarters of the Cherokee tribe. There with a loosely formed company which included ten Cherokee citizens they obtained a lease on 208,000 acres in the northwest corner of the Cherokee reservation. With that lease in hand Keeler interested Guffey and Galey, who had recently gained fame as oil developers in the Kansas fields. They agreed to drill if the signature of Little Star, Secretary of the Cherokee Nation, could be obtained on the lease. A messenger sent out by Keeler chased Little Star all over Texas before he caught up with him and obtained his signature. In the meantime Guffey and Galey's thirty-day time limit agreement had expired and they had moved elsewhere.

Keeler had not forgotten Cudahy's interest in a test well near Muskogee some years earlier. The storekeeper managed to reach the packer's representative and interest him in the oil possibilities on Caney River. A tentative agreement was made for Cudahy to drill a well on each five miles square of the leasehold. But Cudahy was not satisfied with Keeler's Indian lease. The Interior Department had begun to show too much interest in the protection of Indian property rights. If Keeler wanted Cudahy to go to all the expense of moving a rig to Bartlesville and sinking a well, he must give some assurance that the driller would not be thrown off the property by some Federal officeholder. So Keeler journeyed to Washington and put the proposition before Secretary of the Interior Hitchcock. Hitchcock declined to endorse the Indian lease on the larger acreage, but did endorse one on one square mile, including the townsite.

That was enough for Cudahy. At the time Amos Steelsmith and Sam Weaver had just abandoned a test well 1,300 feet deep between Tulsa and Sapulpa. Their drilling outfit was available at Red Fork. Keeler sent fourteen teams from Bartlesville to fetch the equipment. Before the job was completed twenty-six wagons, powered by horses, mules and oxen, were required. Roads, especially winter roads, in Indian territory forty years ago were about as bad as roads could be.

The first task was to get the equipment across the Arkansas River. The stream was low, but running under a crust of ice too thin to hold up the wagons but too thick to permit horses to drag the wagons through it. The first move was to cut a pathway through the ice. That done, the teams were tripled, six horses being put to each wagon, and the ford was crossed. On the opposite bank a miserable group of emigrants with footsore half-starved stock and covered wagons had been held up by the ice-covered stream.

Sadly they watched the progress of the oil equipment across the stream. If they could only swap places and move on toward their southern destination! The transportation boss on the oil-rig job was moved to pity. Roughly he commanded one of the emigrants to hook the extra oil-rig

The discovery well at Bartlesville, now almost a shrine of Oklahoma oil men. Sometimes called the first "commercial" well in Oklahoma, it came in in March, 1897, flowing 150 barrels a day, and making Bartlesville the first oil metropolis of Indian Territory. George B. Keeler, one of its promoters, is pictured beside it.
(Courtesy *The Daily Oklahoman*)

horses ahead of his own sorry nags. The oil men's teams made easy work of the covered wagons. By the time the last load of drilling equipment was on the north side of the river the last emigrant and his wife, his bedraggled children and his wagon had been moved to the south side. The oil teamsters waved a grinning farewell and moved on to greater difficulties.

That journey of sixty-five miles from Red Fork consumed three weary weeks. I made it in an hour and a half, forty years later. In an automobile on a smooth paved highway it was difficult to imagine the depths of mud and difficulties encountered by that first drilling outfit. There were days when the caravan did not move its grub wagon and camp outfit because the progress of the freight teams from dawn to sunset did not justify the move. There was seldom a day when a man perched on top of a load as the new camp was made could not look back upon the precise point which he had left that morning. But they got there. Cam Bloom and A. B. McBride drilled the first well for the Cudahy Oil Company at the north end of the Big Caney bridge at Bartlesville. In March, 1897, the well came in, flowing 150 barrels of oil a day from a depth of 1,400 feet.

The oil industry which was to remodel some sections of the last frontier into a state of civilization unsurpassed in American cities two centuries older was on its way. Bartlesville, convenient to the oil fields of Kansas, was its first metropolis. Six months after the Cudahy company opened the first valuable well the Indian Territory Illuminating Oil Company brought in its first well, the Nellie Johnstone, two miles west of Bartlesville.

Note the word "Illuminating" in the title of that company. Illumination was generally believed to be the chief use of mineral oil, or "rock oil" as it was sometimes called, only forty years ago.

There were hardly more than a score of gasoline-powered automobiles in the United States at the time. Steam, generated by burning coal, was the great source of mechanical power. Oil was more valuable for lubrication than for fuel, and more valuable for illumination than for either of the other uses. Even so it had made a long stride from the day forty-four years earlier when its first publicity in the Indian country was based upon its power as a cure for rheumatism and dropsy. But a far greater stride, beyond the wildest dreams of its producers, was started.

Bartlesville was in the proper strategic position to profit. At the moment of the opening of the first profitable well the town was hardly more than a trading post where Jacob H. Bartles had started a flour mill on the Caney River in 1877. The mill, and the general store which was soon established, attracted cattlemen, Indians and others. Young William Johnstone, originally employed by Bartles, joined George Keeler in the establishment of a store across the river.

All goods had to be hauled from Coffeyville, Kansas, by wagon. Money was scarce on the frontier, but hides, venison, beaver, otter and other pelts with a reliable market in St. Louis were plentiful. Johnstone and Keeler did a profitable business from the start of their venture in 1884, running up to a figure of $75,000 or more each year. They traded coffee, sugar, calico, blankets and hardware for whatever marketable items were offered. They hauled the produce to Coffeyville, shipped to St. Louis and hauled back the supplies of trade.

That could not easily be done with oil. There was not a pipe line in the territory. There was not an oil tank. Every drop of petroleum for trade had to be barreled. There was little profit in that. But there were possibilities, clear and enticing, to businessmen outside the frontier. The first wells had to be capped for lack of market, but Bartles and Porter, of Caney, were pioneers with an eye to the main chance. With associates who promised the

necessary influence and discretion they surveyed a railroad right of way to a junction with the Santa Fe line in
Kansas, established title, and sold it to the railroad. In the
summer of 1899 the first locomotive whistles frightened
the deer from their retreats in the brush beside Caney
River. Tank cars could then handle the oil production at
a profit. The town began to grow as it had never grown
before.

More oil wells! More travelers, seeking oil wells! More
money for development! Hotels for accommodation of
promoters and traders! More stores for the supply of increasing needs of a growing city! Bartlesville was on its
way. Other oil-made cities of Oklahoma were to pass it in
size, wealth and activity, but it is still the headquarters of
the Phillips Petroleum Company and the Indian Territory Illuminating Oil Company, of Frank Phillips and
H. V. Foster—two of the richest and most powerful companies and two of the wealthiest and most influential men
in the oil business of the United States today. Phillips'
company is now capitalized at $375,000,000. Foster, I was
told in Bartlesville by conservative citizens who have
watched his progress for half a lifetime, is one of the half-
dozen wealthiest men in America.

Such were the possibilities when the first flowing oil
well was opened in Indian territory only forty years ago.
Only a few persons recognized those possibilities. Oklahoma Territory and the Indian territory were still the last
frontier, with many of the hardships, most of the incon-
veniences, some of the savagery and nearly all the opportunities of the frontier still before them.

A decade was to pass before oil was to revolutionize the
life and outlook of that frontier. In the meantime there
was other work to be done simultaneously with the development of oil. Great ranches still herded their cattle over
vast areas of Old Oklahoma and the Cherokee Strip.
Towns struggled to become cities. Individuals battled for

a livelihood, for material wealth, for political advancement. Schools and colleges were established and improved. Crime continued to take its toll. It was still tough country, the toughest in the United States.

All the gunplay which was notorious in the early days of white settlement was not confined to professional bad men or celebrating cowboys. Persons of better education and environment, laying claim to higher civilization, occasionally resorted to violence to settle their disputes. I am indebted to Mr. George Rainey of Enid for an illustration of this phase of life upon the frontier.

In the early days of Enid, J. L. Isenberg, publisher of the Enid *Daily Wave,* considered it the part of expediency and good politics to write some scathing editorial comment upon the activities of R. W. Patterson, who had been appointed registrar of the land office through the influence of Hoke Smith of Georgia, then Secretary of the Interior. Isenberg, publishing a Democratic newspaper, expected from the Cleveland administration the proper reward of a deserving Democrat in the form of land-office advertisements. But Patterson, for some reason best known to himself, persistently gave the business to the rival and undeserving Republican paper.

Isenberg's editorial attacks upon Patterson became more and more vitriolic as time went on, until at last they suggested that Patterson was capable of most of the crimes known to man.

That was too many. Patterson armed himself and went in search of Isenberg. They met at a principal corner. A quick quarrel was followed by a blow, and Patterson drew his gun and fired. Isenberg, unwounded, took to his heels. Patterson followed into a near-by store, just as Marshal Williams appeared on the scene and commanded him to halt. But the Georgian's blood was up. He paused only long enough to shoot the marshal through the body, and then pursued the editor toward the alley at the rear. Williams managed to drag himself to a doorway, steady himself against the frame and put a bullet through the head of the land-office registrar.

Registrar and marshal died almost at the same instant. The editor did not stop for news. At the public square he leaped into a hack and ordered the jehu to drive to North Enid, where he boarded a train for Fort Worth, Texas. There he remained until after the funerals of the two victims of frontier methods of journalism.

Such incidents, however, were comparatively few and far between. In general the white settlers in Oklahoma Territory were honest, hard-working, self-respecting pioneers. Most of the gunplay was the work of hilarious cowboys or professional bandits. Among the latter the Dalton gang probably has been given more publicity than any other group active in the years preceding statehood. Al Jennings managed to give himself and his followers a bit of publicity both through their depredations and through an autobiography published serially by a leading magazine many lears later, after Jennings had sufficiently reformed to practice criminal law.

The Dalton gang, however, never reformed. It went out in a blaze of gunfire in Coffeyville, when it attempted to rob two banks simultaneously and was met by fusillades from the more conservative residents. That battle, one of the most sensational in the history of banditry in the West, has been sufficiently publicized.

Sufficient to say that Emmett Dalton, youngest of the notorious brothers, then only twenty years old, was the only one of the group of five outlaws who was not killed. He was severely wounded, captured, convicted and sentenced to life imprisonment. After serving fifteen years he was released, reformed, to become a real-estate agent in Los Angeles. Eventually he followed the Al Jennings precedent writing an autobiographical narrative entitled, *When the Daltons Rode*. There he died, July 13, 1937.

But long before that development the Daltons had quite thoroughly terrorized a considerable area of the last frontier. One incident of train robbery, in which they were believed to specialize, will serve to reveal the methods commonly practiced by train robbers since the first train holdup. The technique never changed in any important detail. Sometimes the gunmen of the gang boarded the blind baggage at a water tank or country station and crawled over the tender to take charge of the situation. Sometimes they swung a red lantern to stop the train in unsettled country between stations. In either instance their subsequent procedure differed but little. The robbery of the Santa Fe out of Arkansas City on June 1, 1892, by the Dalton gang may be taken as typical.

An hour after dark engineer Carl Mack saw a red lantern swinging across the track near Red Rock station, and brought his train to a halt. Two bandits promptly climbed aboard the engine, ordered the fireman to uncouple the train behind the express car and forced the engineer to pull down the track half a mile to the cattle-loading pens where the remainder of the bandit gang waited with horses.

The leader of the outlaw group, believed to be Bob Dalton, supported, according to subsequent evidence, by Grat Dalton, Emmett Dalton, Bill Powers, Dick Broadwell and Bill Doolin, marshaled engineer and fireman back to the express car. There they ordered E. S. Whittlesey, messenger, and John A. Riehl, guard, to open the door. Whittlesey and Riehl refused. The bandits opened fire with rifles through the windows and walls of the car. Whittlesey and Riehl returned the fire, but in the darkness no one was wounded.

The outlaws reasoned with Engineer Mack at the points of their revolvers. He pleaded in vain with the messenger and guard to surrender. Then the fireman was forced to attack the door with a coal pick. When the door was sufficiently wrecked the fireman was forced to crawl in, but the messenger and guard ordered him to one end of the car while they stretched themselves on the floor at the opposite end and shouted that they would shoot the first bandit whose head appeared. The outlaws shouted to the fireman to tell them where the defenders were hidden. When he told them they raked the north end of the car. Sixty bullet holes subsequently were counted, but not a shot found its intended mark.

At last Whittlesey and Riehl acceded to the pleas of the engineer and gave up the unequal battle. Three bandits entered the car and disarmed them, but the express company employes insisted that they could not open the safe, as they did not know the combination. A sledge hammer solved that difficulty. The outlaws cleared out the safe and helped themselves to anything else in the car which took their fancy, including the lunch buckets of the expressmen. They then mounted and disappeared in the darkness. The battle had consumed three-quarters of an hour, during which the passenger cars had stood undisturbed only half a mile away.

The engine and shattered express car were backed up to the train, and pulled it into the next station, where the conductor telegraphed the authorities. U. S. Deputy Marshal Thomas and a posse were soon on the trail, but quickly lost it in the wilds of the Osage country. The precise amount of the loot was never made public. Apparently it was enough to last the gang for four months, for it was four months later that they made their last reckless and disastrous raid on the banks at Coffeyville.

A train robbery on the same line, near what is now Perry, Oklahoma, only a month later, was traced to three other men who were sentenced to two years in the Federal penitentiary at Fort Leavenworth. A two-year term might indicate that the frontiersmen did not take train robbery too seriously. The technique in that incident was the same as pursued by the Daltons except that they climbed aboard the engine at a water tank instead of flagging it, and used a stick of dynamite to open the express door. In a dozen train robberies in several states through those same years there was never any important change of method.

But there was a change in environment, in conditions which attracted criminals to the wilds which offered refuge from pursuit. The railroads, banks and stores upon which they preyed were bringing a larger and more law-abiding population into the frontier country. More towns sprang up. The distance between towns shortened. Local officers and courts were established throughout a region many miles in extent which formerly had depended almost entirely upon United States marshals and the far distant court at Fort Smith.

With the discovery of oil in profitable quantities, eastern capital moved into the territory to start new enterprises. Capital must and would protect itself. Capital and politics co-operated, for better or for worse. Armed banditry was doomed. A state was in the making.

CHAPTER XI

OIL VS. CATTLE

CATTLE by the thousands still roamed vast areas of Oklahoma and Indian territory when the first real oil excitement struck the frontier. Wheat and corn had made some settlers prosperous, but the cattle business was the source of romance and color, of riches for a favored few and livelihood for thousands.

The 101 Ranch of the Miller Brothers, because of the Wild West Show and circus which it eventually promoted, probably has been the most widely publicized cattle ranch of the territory. Other ranches such as the Chapman and Barnard ranch of eighty-five thousand acres, or the show ranch and game preserve of Frank Phillips, have become famous. But they have been financed by oil rather than by the cattle which they produce, although they are now profitable in their own right.

More interesting, in its revelation of the flood of oil as opposed to cattle, rather than as source of capital for the cattle business, was the "3-D" ranch, a range of sixty thousand acres in the Osage Nation. Lon R. Stansbery, still living in Tulsa, familiar with that ranch for half a century, has written a vivid account of that losing battle of frontier life against oil.

"We old timers think of its wonderful prairies, its magnificent hills, and the creeks of cool sparkling water filled with perch, crappie and black bass; but today its plains are dotted with oil derricks, the grass is covered with oil, the beautiful streams are filled with salt water and fish have long since been strangled out. The black walnuts, the beautiful elms, redbuds and cottonwoods are all dead, and the great bluffs and rocks are now black with oil. The pumphouse stands on the old round-up grounds. The old 'snubbin' post,' where many a bronco first felt the cowboy's rope and began to learn the cow business, is used by teamsters with a block and tackle attached for pulling wells. The horse pasture is dotted with roustabout shacks and a supply house now covers the site of the old corral and branding pens. Where they used to stack barrels of salt you now find stacks of eight and ten-inch pipe. That is what I mean by 'The Passing of the 3-D Ranch.' "

Students of frontier life are indebted to Lon Stansbery for more than that picture of the physical change which oil wrought upon the last frontier. His *Passing of the 3-D Ranch*, written and privately printed for the edification of a few old friends, is a vivid reflection of the frontier life and the humorous philosophy of the frontiersmen which were to give way to the conventional forms of civilization financed by the wealth of oil.

Stansbery explains that when he came from Missouri to the little town of Tulsa in 1889, a clothing salesman in Coffeyville, Kansas, outfitted him with an overcoat "so long that it drug out my tracks and I could never find my way back to Missouri." There were only three stores in the village, according to Stansbery, one selling groceries, one selling firearms and the third selling Peruna. For the information of a younger generation, Peruna was a patent medicine of high alcoholic content, much in demand in Indian territory where the sale of whiskey and similar beverages was restricted.

The year was that of the opening of the Unassigned Lands in central Oklahoma to the first great land rush. In preparation for that opening the cattlemen who were grazing large herds on Creek and Seminole lands had been ordered to remove their stock.

Tom Wagoner of Texas, owner of some of the largest herds in the country, leased grazing rights at three cents an acre on some sixty thousand acres of land in the Osage Nation, and drove fifteen thousand head of cattle across the Arkansas river to start the "3-D" ranch. Two years later Stansbery left school in the village of Tulsa to become "salt-boy" on the "3-D." In those days in that area "salt-boy" was the first step in apprenticeship to the heights of youthful frontier ambition—"top hand cowboy, peeler, waddie, buckaroo," and so forth.

Lon's teachers in the arts of range-riding, rope and branding iron included some of the real aristocrats of the range. Bill Doolin, whom Stansbery remembers as one of the best cowhands in the country before he turned outlaw and was killed, was one of them. The salt-boy's job was to haul salt by the barrel, five barrels to a wagonload, from J. M. Hall's store in Tulsa to the "3-D" range, and distribute it to the "licks" for the cattle. Between times he was the butt of practical jokes by the cowboys, and picked up whatever valuable experience and information he might.

Nearly half a century later Stansbery explained his change from youthful ranch hand to businessman as being due to one of those practical jokes. The cowboys of the "3-D" had saddled and "double-geared" a bronco which he was to ride to Tulsa to carry a special message. As soon as the boy had mounted, the horse began to buck.

"After a third or fourth jump I decided I could not ride him straight up cowboy fashion and not pull leather. I reached for the saddle horn, but my arm was too short. The Arkansas river was twelve miles from the ranch, but I went up so high I remember seeing the water, and the bluebirds built nests in my hip-pockets before I came down. This little escapade probably spoiled a good cowboy, as it took away my desire ever to be one."

It did not, however, take away Stansbery's love of the open range, or a sense of humor which has been associated with cowboy lore since the first loop was thrown. Ex, amples of that humor and an unspoiled originality suggestive of the late Will Rogers make his reminiscences a delight. They are also, I believe, an accurate reflection of conditions and character on what remained of the old frontier after the homesteaders, sooners and nesters had established themselves, and before oil completed the change.

Real-estate and business conditions in Tulsa at the time are revealed in his statement that Berry Hogan offered him a lot where the Exchange National Bank, the city's largest skyscraper, now stands, if the young man would cut the weeds on it. "But I was so busy sowing wild oats that I had no more use for a lot on Third street than Noah had for a tail light on his Ark."

After a few years of varied experience in and around Tulsa, Stansbery moved to Davis to manage a dry goods store, but was so homesick that he "was in more misery than if Geronimo had staked me out on an ant hill." Safely back in Tulsa, he decided that he should marry and settle down. So, he says, he "advertised for a wife. I got answers from eighteen different Old Timers saying I could have theirs."

Social contrasts between the past and present are covered in three sentences: "Some of the boys in high-powered cars think they are going pretty fast, but I want to say that some of us Old Timers could put over a few things with a horse and buggy. You could buy your girl a glass of lemonade for five cents, then drive along at the rate of two miles per hour. The records show that when you became engaged driving along at this rate and got married, it lasted longer than when you became engaged at the rate of thirty-five' or fifty miles per hour."

"We have had lots of good men come to Tulsa and look it over, but it did not seem to have any more attraction for them than a short skirt would have for Adam. . . .

"I met a man on the street the other day who had only been here about three years. He asked me if I had ever been on the stage. I told him, 'No, but I had an uncle who had been on the stage.' He took the part of the drunkard in 'Ten Nights in a Bar Room,' but only lasted six months as he killed himself in rehearsing. However, I had a niece who went to Hollywood in the early days and certainly made good in the movies. She ran for the doctor in 'The Birth of a Nation.' "

Commenting on the small beginning of Tulsa, Stansbery remarks that Harry Campbell, later a leading jurist and president of the Pioneer Association, arrived in town with one law book, and Dr. Hawley opened his office with three pills.

"We had only two or three doctors in town in the early days and about all the business they had was chills and fever, but they finally got the idea of operating. . . . They operated on Jack Cleary so many times he finally asked them to put a swinging door in him." When Stansbery himself in later years was ordered to Hot Springs, Arkansas, for a course of treatments, he says the first doctor he met offered to examine him for $100. "I told him to go ahead and if he found it I would give him half."

Describing a fight which broke up a meeting of residents considering a town problem, Stansbery reports that "Lon Conway hit Roxie Roberts so hard his shirt tail rolled up like a window shade and almost choked him to death before anyone could get to him."

As an example of real-estate opportunities he recalls that when M. J. Glass, an old-time cowboy and later a member of the Miller & Glass Real Estate Company, arrived in Tulsa he heard Dad Lennox, hack driver at the station, shouting, "Any part of the city for fifty cents." "M. J. rushed over and handed him fifty cents, saying, 'I'll take the north half.' This of course was a mistake on M. J.'s part as the south half was a much better buy."

Stansbery cites in modest evidence of his own practical shrewdness an occasion when he was called upon to address a Real Estate Men's banquet at the modern Mayo Hotel. His only previous experience of the kind, he says, was when he addressed an Anti-Horse-Thief meeting on Bird creek in 1900.

"After I knew I was going to make this talk I called on over thirty real estate men and told each one I was interested in a certain piece of property, asking him to get me a price on it but to not let anyone know I had made an inquiry. The next evening every time I told a joke those thirty men led the applause and put me over."

But, in common with many of the cheerful-souled pioneers who have observed and assisted in the development of the crude frontier to its present state of civilization, Stansbery is a sentimentalist as well as a practical philosopher. He can never forget nor quite forgive oil's destruction of the beauties and advantages of the natural countryside.

The beginning of that change had come with the opening of the first potentially profitable well at Bartlesville. Two years were to pass with little change upon the frontier before the possibilities of profit brought the first railroad into the first tiny oil metropolis. Tiny it was, but far more significant than any pioneer of the frontier could imagine.

"At Pawhuska, in the Osage Nation," says an extract from the Oklahoma Governor's report to the Secretary of the Interior in 1897, "two test wells have produced petroleum of superior grade, while there are many indications of oil at other places in the reservation."

In the following year the Governor's report revealed the first suspicion that leaders in the nation's oil business were preparing for a coup as soon as they could obtain the needful authority. "A well put down at Pawhuska a year ago," says the report, "struck some oil but was immediately sealed up, the drillers declaring that the yield was too small to make it pay commercially. It is a significant fact, however, that the company that put down this well afterwards leased large tracts of land in the Osage Reservation and in Pawnee and Payne counties. At Bartlesville, just over the line in Indian territory, oil of an excellent quality has recently been struck in paying quantities, and at Muskogee and Eufaula, in the Creek Nation, several paying wells have recently been drilled. . ."

Belief that oil underlay the surface throughout a far larger section of the country was indicated in the following year when the Governor's report said: "There are unmistakable evidences of oil and gas at many points in the territory. In Payne and Pawnee counties are several springs where the water is polluted with oil, and in the Chickasaw Nation and the Kiowa and Comanche Reservations are oil springs. . . . Paying wells are being put down in all parts of the Creek country, just east of Oklahoma. A local company is putting down a well at Guthrie. . . ."

Still there was no recognition of conditions which were to play history's most ironic practical joke upon a government which had established the Osage Nation upon what appeared to be the most useless land in all the Indian country. No one suspected the extent or poetic justice of that forthcoming jest. There was no rush, no great boom, no great excitement. Probably that momentary respite to the doomed life of the last frontier was due chiefly to the restrictions upon Indian lands. Promoters realized the advisability of laying their political and economic lines so that they might hold and work the ground after having leased it from the Indians.

That was the situation in 1901 when Dr. J. C. W. Bland, a practicing physician and an intermarried Indian citizen in the commonplace little cow-town of Tulsa, noticed a drilling rig on a siding of the Frisco line at Red Fork, four miles from Tulsa. The result was the famous Sue A. Bland well at Red Fork, which many oil men still declare was the inspiration of the first real oil boom in Oklahoma. Details of events leading up to that boom are still the subject of heated controversy among old-timers. One faction gives all the credit to Dr. Bland and his friend and associate, Dr. Fred S. Clinton. Another faction accords the honor to J. S. Wick and Jesse A. Heydrick, oil promoters from Butler, Pennsylvania.

The facts, as accurately as I have been able to deduce them from conflicting testimony and some documentary evidence, seem to be as follows: John S. Wick & Company first obtained a lease, recorded December 14, 1899, from a group of Indians including Thomas J. Adams, Wash Adams, Lewis Adams, James Sapulpa, William Sapulpa and a few others. Because of the Interior Department's restrictions on the leasing of Indian lands, and the fact that allotment of those lands was then being made by the Dawes Commission, there was some doubt as to the legality of that lease.

At the moment it was not put to a test. It lapsed automatically because of a six months clause which required the start of drilling. In the following year the same group of Indians, with Sue A. Bland, the doctor's wife, included, renewed the lease to Wick & Company. Still there was no action. Wick's loud claim that he was a cousin of Andrew Carnegie and a brother-in-law of Secretary Hitchcock, and that therefore any lease which was made to him would be honored, did not impress oil men sufficiently to finance him very extensively.

It was not until the spring of 1901 that Perry L. Crossman and Luther Crossman, drillers, were ordered to ship a rig from Joplin, Missouri, to Red Fork. Attempting some years later to settle the controversy as to initial responsibility for the Sue A. Bland well, Perry L. Crossman wrote to James C. Heydrick, a son of Jesse Heydrick, saying that he made a personal contract with Jesse Heydrick to drill the first well at Red Fork, and that John Wick's name was not on the contract.

In any event, it was the Crossman rig, held up for lack of payment of freight charges, that caught the eye of Dr. Bland. One story is that the Frisco agent had refused to accept a draft tendered in payment of the charges, and that Bland and Clinton cashed the draft and each was given one one-hundred dollar share of stock in the enterprise. Neither Wick nor Jesse Heydrick is living to give his side of the story.

Thirty-six years after the incident Dr. Clinton, retired from practice because of ill health, in his home in Tulsa told me his version of the controversy. He asserted that Wick and Heydrick had obtained an oil lease on five hundred thousand acres, approved by the Creek Council, which consisted of a House of Kings, a House of Warriors, and an elected Chief. But the lease was not approved by the Secretary of the Interior.

The drilling rig which had aroused Dr. Bland's interest, according to Dr. Clinton, was actually on its way to what a few years later was to become famous as the Glenn Pool, but was held up at Red Fork by a question as to the financial responsibility of the consignee. It was diverted by Bland and Clinton when the latter borrowed three hundred dollars from the Frisco agent on a note at ten per cent interest to pay the freight and provide a grubstake for the drilling crew.

Perhaps Crossman did not care much where he drilled or who paid him for it, as long as he was paid. That point has not been settled. In any event, a site for the well was selected just outside the surveyed townsite of Red Fork because of the fact that the townsite was not open to allotment. Neither was the selected site, as commonly asserted, already allotted to Sue A. Bland. That came later, in a twenty-four-hour period of activity remindful in miniature of some of the world's greatest gold-rushes.

Drilling began on May 10, 1901, under the superintendency of Heydrick, with Perry L. Crossman in direct charge of the work. Dr. Clinton asserts that he financed the job to the extent of providing a grubstake for the workers, and that both he and Dr. Bland kept close watch on developments. On the night of June 24, 1901, while both Heydrick and Crossman were absent, the drillers encountered gas. The next morning the oil came in with a rush from a depth of less than six hundred feet—the first real gusher in Oklahoma.

Dr. Clinton was on the scene, as excited as a young man could be with a fortune poured at his feet. When Wick appeared he excitedly wired Perry Crossman in Joplin: "Send packer; oil is spouting over top of derrick."

But Dr. Clinton was aware of the possibilities of greater loss than the immediate loss of oil. Dr. Bland was ill with appendicitis. Clinton hurried to tell him of the strike, taking along a lawyer and a notary. As quickly as the papers could be drawn and certified Dr. Clinton obtained Sue A. Bland's power of attorney. Equally important, in his opinion, he had a quart bottle of the oil. That fact is an interesting sidelight upon how little the physician, or the majority of the persons in the territory at that time, knew about oil. Its genuineness still had to be proved to them just as the first gold picked up from Sutter's mill race had to be proved to James Marshall.

Dr. Clinton was quite certain that this was oil. Its smears upon his hands and clothing, liberally acquired from the spraying gusher while he filled a bottle with the fluid, left little doubt of that. But there was in his mind some question of its quality and value, and of his ability to convince an important authority that oil existed where no oil had ever been suspected.

It was a long way to Muskogee, and in Muskogee, seat of the Dawes Commission, lay the machinery necessary to turn this adventure into a fortune. The first and smartest man to arrive there with the needful credentials and evidence could do it. Dr. Clinton intended to be that man. Without stopping to change his oil-soaked clothes he boarded the first train out of Tulsa for Vinita and made connections with a train running to the Indian capital. It was late at night when he arrived and hurried to the home of Dr. F. B. Fite, a personal friend and a man of influence in the Dawes Commission.

Dr. Fite scrambled out of bed as country doctors were in the habit of doing on midnight calls. Doubtless he was relieved to learn that he was not expected to drive a score of miles into the country to officiate at some birth or death. Dr. Clinton poured forth his story, and offered his evidence. The two doctors adjourned to the back yard to test the oil. They were not certain that it would burn. If it did burn it might blow up in their faces. They poured some on shavings. It burned gaily. Taking courage in their hands, they filled a tin lantern with the fluid and put in a new dry wick. It flared to the match like kerosene. It was oil, and oil of superior grade. What next?

Dr. Fite routed his friend Secretary Aylesworth of the Dawes Commission out of bed, explained the situation and arranged to have the necessary papers ready at the office of the commission at eight o'clock in the morning. Dr. Clinton presented his power of attorney for Mrs. Bland, and the allotment of forty acres surrounding the gusher was made and officially recorded. Weary but triumphant, Dr. Clinton made his way back to Tulsa and Red Fork.

Half of Tulsa was standing around gaping at the spouting oil well. But half of Tulsa was not much in 1901. Neither was the well much of a gusher compared with some which were to make the Oklahoma oil fields famous in later years. But it was exciting. Fred S. Barde, corres-

pondent of a leading Kansas City newspaper, gave the
news its first wide circulation. A dozen other newsmen
appeared and helped. In another day the rush was on.
Twenty-five hundred persons were on the scene in another forty-eight hours, coming by wagon and horseback
from nearby towns, and by rail from such distant points
as Independence, Kansas, and other oil-conscious regions.
Red Fork had no accommodations. Tulsa was filled beyond overflowing. From that moment Tulsa's faith in oil
as the source and sustenance of civic grandeur never faltered.

The well was flowing one hundred barrels a day. Oil
was worth a dollar a barrel at the moment. It appeared
to be big money until the cost of transportation by rail
to the nearest refinery, in Kansas, at ninety cents a barrel, was deducted. That cooled the ardor a bit. Still the
price of town lots, in the area free of the Federal restrictions on Indian lands, and therefore open to drilling,
jumped from fifteen cents to fifteen dollars a front foot.

An offer of $40,000 cash for the well and Sue Bland's
forty acres around it was refused, "because of no sense,"
says Dr. Clinton. Dr. Bland and Dr. Clinton found themselves enjoying unaccustomed financial credit, and proceeded to buy what they called "deeds of possession" to
neighboring allotments. Mrs. Bland finally sold a lease
to Oscar R. Howard at considerably less than the original
offer.

Various persons, local and otherwise, loaded up with
leases which they were not equipped to develop. The
boom, as such, collapsed. But interest in the oil business
had been highly and permanently stimulated. Tulsa had
enjoyed its first taste of oil and found it a habit-forming
drug. Its excitement and profits in caring for the hundreds who rushed to Red Fork made the days of its de-
pendence upon cattle and shipping seem dull indeed.

One of Oklahoma's men of energy, ambition and

Above: Tulsa as it appeared in 1890.

Center: Tulsa's Main Street in 1909, definitely launched on its way to becoming the oil capital of the world.   (Courtesy Wyatt Tate Brady Collection)

Below: Panorama of Tulsa in 1937, revealing one of the most impressive groups of skyscrapers in the United States, all built on the oil business.

imagination was Robert Galbreath. With energy somewhat impaired by the passing of thirty-odd years, but with
ambition and imagination still quite evident, Mr. Galbreath admitted me to his home in Tulsa and told me
something of his experience.

He did not tell it all, by any means. He appeared to be
a bit suspicious. As the story unfolded under careful
questioning the reasons for the suspicion unfolded with
it. Galbreath had had numerous unsatisfactory experiences in more than thirty years of the oil business. He
had been in the big money, and out of it. He had defied
some of the most powerful interests in the oil trade of
the United States, and had not forgotten, or forgiven, or
been chastened by the results.

As he described the development of those early days,
"government was by telegraph," or "carpetbag rule." By
that he meant that almost every deal of importance with
reference to oil rights on lands allotted to the Indians, or
held by the tribes awaiting allotment by the Dawes Commission, was subject to extensive telegraphic negotiations with the Secretary of the Interior. And the well-
financed men who came into the region with the hope of
seizing and controlling and exploiting its oil resources
too frequently appeared to have the best telegraphic con-
tacts with Washington.

In the course of time they cost Galbreath a lot of money.
But they never broke his spirit, his sense of right and
justice or his growing enmity toward what appeared to
be entrenched privilege. Very early in the game he became suspicious of them. He is still suspicious of strangers.
He was suspicious of me. Also he had been thinking of
writing a book. Why should he give away facts and information of which he alone was aware, and which might
net him cash profit if properly presented under copy-
right?

That was familiar ground to one who had interviewed

a hundred men in research for half a dozen books. I have met old pioneer women in the hinterlands of Montana who were going to write a book about the exciting days of the copper wars, and live out their remaining days in comfort on the royalties. What a professional writer could tell them about royalties! I have met country storekeepers, the sons of pioneers, who guarded and hid the reminiscences of their stage-driving fathers as gems beyond value, because they were going to write a book. I have encountered one librarian in a backward state, elevated to the position by the political influence of friends of a pioneer father, who was cold to all questioning because her brother was going to write a book. Many of my questions concerning the past have been turned aside by plumbers, waiters, bartenders and crippled miners because they were going to write a book about that themselves when they got around to it.

So I told Robert Galbreath that I would await his book with interest, when, as, and if issued. In the meantime I must gather such distorted information as I could from sources less closely connected with the phenomenal development of eastern Oklahoma than himself. The Prairie Gas & Oil Company, I suggested, had left some men and some records still available which shed considerable light upon the advantages of organized capital's interest in oil development.

"Huh!" said Mr. Galbreath. "Jim O'Neil of the Prairie cut the price of oil almost in half, to fifty cents a barrel, as soon as my first good well came in at Red Fork. That was in 1904."

"Red Fork? I thought Bland and Clinton opened the first well at Red Fork, or Wick and Heydrick, if you prefer that side of the controversy."

"That was two years earlier. That first boom just gave us a touch of the oil fever. I got hold of some property adjoining the Bland allotment and interested Charley

Colcord and C. G. Jones, mayor of Oklahoma City. It took some time: Local government here in Indian territory, as I told you, was by telegraph. Everything had to be put up to Secretary Hitchcock's department in Washington.

"We started to drill on some land we leased from John Yargee west of Red Fork in 1902. We didn't have Secretary Hitchcock's approval, and pretty soon along came the Indian police, departmental sleuths, carpetbag scouts and uniformed bands of brass collars, forbidding and warning against further drilling. It looked like the penitentiary for us, but we managed to keep out by working under a twelve-months occupancy feature of the homestead regulations which seemed to give Yargee the right to lease.

"We continued to drill and brought in a ten-barrel producer. Other wells were completed. We managed to keep out of jail and to continue blazing the trail of Oklahoma's oil industry. Finally, after a big battle, the government gave approval of certain leases, including our John Yargee property. Immediately afterward we found a forty-barrel well at twelve hundred feet. In February, 1904, our No. 3 Nathaniel Yargee came in for one hundred and twenty-five barrels. It was completion of this well that extended important production northeast of Red Fork and brought Jimmie O'Neil, president of the Prairie Oil and Gas, and other officials of the Prairie down from Independence, Kansas.

"I unloaded my enthusiasm to O'Neil. I fairly bubbled over with my predictions of a vast petroleum production in the Red Fork and 'Mounds' district to the south. I was right about the latter too, as the Glenn Pool proved. O'Neil did not do a thing but go back to Independence and cut the price of crude oil to fifty cents a barrel. That was a case where a one hundred and twenty-five-barrel well broke the petroleum market."

Evidently there was considerable discrepancy between the market for oil and the popular interest in oil little more than thirty years ago. It was still looked upon more as a source of illumination than as a source of power. But R. E. Olds had recently established the first quantity-production automobile factory in the world with an output of 425 cars in 1901. Henry Ford was obtaining the experience necessary to the launching of the Ford Motor Company with a first year's production of something more than 1,000 cars in 1903. John D. Rockefeller and the Standard Oil Company were demonstrating values and profits. The potential demand was ready to grow with the output.

In the meantime Frank Chesley, in business in Keystone, had become associated with Galbreath. Chesley was interested in two small wells in the Red Fork area which Galbreath was pumping with so little profit that he announced his intention to quit.

Robert Glenn, a farmer living a few miles to the south on land held as an allotment to his wife, a one-eighth Creek Indian, induced Galbreath to examine a limestone ledge which was heavily stained with oil seepage, some hollows in the stone actually being filled with a translucent green oil. Chesley joined Galbreath in a proposition to drill on the Glenn allotment. It was a wildcat operation, ten miles from the nearest producing well. To reduce initial expense to a minimum it was drilled without a casing, but oil came in with a rush which sprayed over the top of the derrick. The date was November 22, 1905.

The Glenn Pool was discovered. The future of Tulsa was hardly in doubt after that. Galbreath had profited by earlier experience and observation at Red Fork, and had no intention of allowing this fortune to get away from him. For a time an armed guard kept wayfarers away from the scene while Galbreath went about the task of financing the enterprise without publicizing his discovery and thereby handicapping his efforts to obtain leases on surrounding areas.

He sold some West Tulsa property to John O. Mitchell for a note for $1000 secured by a mortgage. After some difficulty he succeeded in discounting the note at the local bank which had opened at Red Fork. That money was used to case the well, and reveal a regular daily production of 125 barrels. It quickly came to fame as the Ida Glenn No. 1. Colcord and Mitchell came in with $5,000, secured by an interest in 870 acres of oil leases which Galbreath and Chesley had obtained in the neighborhood. The money was urgently needed to pay bills incurred in drilling and lease-buying.

A second hole was dry. A third produced 1000 barrels a day. The next brought in 2500 barrels daily. The "big money" interests began to rush for leases and to put pressure on Galbreath and his associates. Galbreath's hostility to anything carrying the brand of Standard Oil or any of its subsidiaries developed rapidly as he was forced to run his oil into earth-tank storage because of a lack of pipe-line facilities, steel storage, and market demand. That hostility was to be an outstanding feature of his character through the years to come. His associates were not so particular as to the identity of a prospective buyer of their interests. They sold for $1,500,000. Galbreath held on until J. E. Crosbie paid him $500,000 for his holdings.

At that, Galbreath did better at the moment than Colonel H. L. Wood, who wrote a slightly different story of the Glenn Pool which was published in the *Oil Trade Journal* twenty years later. Galbreath reports one dry hole, the second, among the first 135 wells drilled in that area. One paragraph of Woods' story reads as follows:

"Kiefer, metropolis of Glenn Pool, was a two-car siding when I scouted the first two dry holes Bob Galbreath worked off on an Independence, Kansas, banker for $2,000, and he sold to friends for $3,000. The big oil man I staked to a 16-acre lease from which he sold $4,000,000 worth of crude oil and casinghead gas gave me a drink of Scotch two years afterward, which is about the biggest commission ever paid me."

It is interesting to note that the greatest casinghead gas development of the region was under the direction of the same J. E. Crosbie to whom Galbreath eventually sold his holdings. Crosbie, a Canadian Scot, and one of the unique characters in the oil history of Oklahoma, had come into the Glenn Pool in the service of the Gypsy Oil Company, controlled by the Mellon interests of Pittsburgh. But Crosbie is a dour Scot. Probably he is not the one who gave Colonel Woods a drink of Scotch in consideration of a $4,000,000 profit.

Crosbie learned his oil drilling in India. Apparently he learned it well, and already possessed the more valuable characteristics of a promoter and banker. The Gulf Pipe Line Company, a Mellon promotion, offered him his opportunity when its Gypsy Oil Company gave him a quarter interest in two leases in the Glenn Pool to develop production which would justify a line to the gulf. The Texas Company laid a line from Port Arthur, Texas, at approximately the same time, but the Gypsy obtained the cream of the field. Crosbie took his share.

He bought one lease from C. C. Drew for $37,000, obtained the Galbreath holdings for $500,000, and otherwise expanded his personal holdings until at one time his income was popularly reckoned at $800,000 a day. Not many days of that were needed. Fortunately for Mr. Crosbie it was long before the days of Federal income taxes. He bought controlling interest in the Central National Bank of Tulsa in 1908 and became its president. In eighteen years the bank's capital increased from $100,000 to $1,000,000, and its deposits from $389,000 to $10,000,000.

When I made an effort to ask him about that $800,000 a day I was refused admittance. Perhaps the chances would have been a little better if I had inquired about his trotting-horses. For more than thirty years Mr. Crosbie has entertained himself with big business and relaxed only with the breeding and racing of trotting-horses. He has long been a familiar figure at the most important harness-racing meets in the United States and Canada. He has no interest in the more spectacular "sport of kings." He never bets. He lives frugally. He is contemptuous of display.

He was one of numerous tremendously successful men in Oklahoma's oil history who obtained their start and many of their millions in the Glenn Pool. Among those who will appear later in this narrative were Charles Page, Robert McFarlin, and James A. Chapman. Of another type, eventually a failure but far more spectacular in the days of his success, was Billy Roesser.

But enough, for the moment, of the Red Fork and Glenn Pool discoveries and their original stimulation of Tulsa. There were other factors of equal or greater importance going on in the Indian lands around the earlier oil metropolis of Bartlesville.

CHAPTER XII

Indians Have the Last Laugh

It should be remembered that the so-called Indian territory was still under control of the Federal government, without votes or elected representation in Congress, when the first oil well came in at Bartlesville. At the same time the Osage Nation, owning in common an area larger than the state of Delaware, maintained its tribal integrity within the territory of Oklahoma.

That was the situation in the early nineties when Edwin B. Foster and Henry Foster, brothers, came west from their home in Westerly, Rhode Island. They were no ordinary farm-hungry pioneers. They were looking for investments to increase a fortune which was already fairly large for those days. A third brother, Barclay Foster, remained at home to look after the family interests which included control of the Mechanics Savings Bank of Westerly.

The Fosters promptly became interested in oil development which was starting in the Kansas field. John N. Florer, an Indian trader who had lived among the Osages since their removal from Kansas to the rocky lands east of the Arkansas in 1871, interested the Fosters in a possible lease of oil rights covering the entire Osage country. On March 16, 1896, a lease was signed by James Bigheart, Principal Chief of the Osage tribe. It gave the Fosters exclusive right to prospect for petroleum throughout the Osage reservation and produce it on a basis of royalties to be distributed among all members of the tribe.

The lease was promptly approved by Hoke Smith, Secretary of the Interior in President Cleveland's cabinet.

It was an approval later to be characterized by another Secretary of the Interior, Ethan Allen Hitchcock, as "the scandal of the time." Certainly it was productive of scandal, crime and ruthless exploitation worthy of the last frontier. Earlier leases had been signed by the Indians, but the Osage lease to Edwin B. Foster was the first to have been officially approved in Washington, and the greatest in area involved that has ever been recorded.

Henry Foster died before the lease was approved. Edwin B. Foster organized the Phoenix Oil Company and transferred the lease to it for immediate action. One of its clauses provided that work should be carried on with no periods of delay longer than six months. Evidently the Indians had had some wise counsel on the wording of the document. James S. Glenn, who immediately joined the Phoenix company, proceeded to drill at a point south of Chautauqua Springs, Kansas. The well was a duster. Nos. 2, 3 and 4, sunk at various places in the reservation, were unproductive.

The exploration was running into heavy expense. Transportation was difficult and costly. Some drilling rigs were brought from as far away as Texas, and hauled for long distances by ox and mule teams over mud roads which were almost impassable to heavy wagons. The Foster fortune, largest single source of financial backing of the Phoenix company, began to feel the strain. No stock jobbing system of financing oil prospecting had been perfected. Oil promoters, and even an oil lease covering such a vast area as the Osage reservation, were not looked upon with favor by bankers in general. The Mechanics Savings Bank of distant Westerly was imperiled. Edwin B. Foster and Barclay Foster followed Henry to the grave.

The Osage Oil Company and Samuel Sheffield took over part of the blanket lease. By 1901 almost the entire Foster fortune appeared to have been sunk. But H. V.

Foster, of the next generation, was still a potential asset. He joined John Brennan, counsel for the Foster interests, and the Indian Territory Illuminating Oil Company was organized to succeed the Phoenix. Throughout the next thirty-five years it was to be one of the greatest names in the oil history of Oklahoma. Every creditor was given stock in the company. Those who retained their interests eventually had ample reason to be thankful that their demands had not been met with cash.

Still the I. T. I. O. had little money, though it held lease rights on vast areas of land. By that time the first wells in the Bartlesville area at the edge of the Osage Nation had proved themselves. The Sue A. Bland well at Red Fork had come in. Geological study extending southward from the Kansas fields presented convincing testimony that the Osage, the Cherokee, the Creek and other lands should be rich in oil. Independents were ready to take a chance. The I. T. I. O. had little difficulty in finding companies or drilling contractors who would sublease rights and proceed to drill. Development in the Osage country accelerated. In the next three years oil production from Osage lands grew from 20,000 barrels to 641,000 barrels.

The greatest, most ironic practical joke ever played by Fate upon the government and taxpayers of the United States was under way. The Federal authority which had been stripping the Osages of their lands from time to time through almost a century found itself at last giving rather than receiving. Pause for a moment to review the history of the Osage tribe.

When it first came to the attention of traders, trappers and the Federal authority a century before the initial opening of the mid-continent oil fields, the Osage tribe roamed and claimed most of what is now Arkansas, Missouri, Kansas and Oklahoma, with some areas to the northward and some east of the Mississippi river. By treaties in 1808 and 1819 they were forced or induced to cede large areas to the encroaching whites. In 1825 they were induced to cede a strip through their remaining lands for the original Santa Fe trail for a paltry $500. In the same year they relinquished all claims to Missouri and Arkansas and all lands to the westward south of the Kansas river. Their lands were then officially designated as an area fifty miles wide extending westward from a line twenty-five miles west of Missouri to the limits of U. S. jurisdiction. In that area they lived until 1865, when 1,500 square miles on the eastern end of their territory were taken by the government with a payment of only $3,000 to the tribe. At the same time an area twenty-five miles wide along the entire northern edge of their country, comprising half of its entire area, was taken over by the government with a promise that when it had been sold to settlers at $1.25 an acre the money so received would be held in trust and used for the benefit of the Osages. Five years later the last of their lands were turned over to the state of Kansas and the remnants of the tribe was moved upon the apparently worthless acreage which is still their homeland.

For a century the Osage tribe had been pushed along the downgrade by the United States government. And, by the irony of Fate, from the depths of its barren lands and its tribal degradation it drew a prize which was to make it the richest nation in the world.

In the year 1925 alone, for example, every individual of the Osage Nation with originally assigned or inherited rights in the nation's land was to draw a minimum of $13,200 from the oil which had never been suspected when the tribe was assigned to those rocky pastures.

With a reduction of nearly fifty per cent in the actual numbers as counted and recorded for allotment of equal rights in the tribe's oil lands in 1906, many of those remaining had inherited numerous additional shares, or headrights, as they were called. One Osage in 1927 was to draw $112,000 as his annual income. The poorest of the entire nation received more than $1,000 a month.

Who should throw the first stone, or the first laugh, when one squaw, relieved of the restriction of a guardian, fifty years after her ancestors had sold 1,500 square miles of their land to the United States for $3,000 promptly used her first unrestricted oil royalties to buy a $1,200 fur coat, a $3,000 diamond ring, pay a $4,000 installment on a California house, purchase $7,000 worth of furniture, lend $1,500 to a sister and put $12,000 into Florida boom land? Who should sneer at a blanketed buck and squaw, whose parents had eked out an existence on government rations, when they appeared in an Oklahoma store, bought eighteen dollars' worth of spring strawberries, squatted on the sidewalk and consumed the lot with gusto? But more of such incidents in their place.

The Osages as a nation, within their intellectual limitations, returned good for evil to the government which had despoiled their forebears. When the United States entered the World War their tribal council gave five thousand acres of oil land for a naval oil reserve. Individuals of the tribe bought $2,500,000 worth of Liberty bonds. Many contributed liberally to the Red Cross. One-third of their eligible men, not subject to the draft, volunteered for military service.

Some years prior to that development, however, the original Foster blanket lease on the Osage lands had encountered not only difficulty in finding oil but other difficulties engendered by competition when it did find oil. When the original ten-year lease on 1,500,000 acres expired in 1905, more than half the acreage was withdrawn from lease. Renewals were made chiefly to Indian Territory Illuminating Oil Company, the Foster promotion and sub-lessees, operating over a field of 680,000 acres along the eastern side of the reservation. But production jumped from 641,000 barrels to 3,400,000 barrels in a single year.

In the meantime numerous factors had entered the situation. The Curtis Act, passed in 1898, had provided for allotment of land to each individual member of the Five Civilized Tribes, the abolition of tribal courts, the discontinuation of tribal governments after eight years. Additional powers had been conferred upon the President and through him upon the Secretary of the Interior to safeguard the rights and property of the Indians, particularly the Osage Nation which was to continue to hold all its land in common for the benefit of all its people. Allotment of surface rights to Choctaws and Chickasaws were not to include mineral rights. The latter were to be leased by the government and royalties used for education and other benefits to the tribesmen. All such exceptions were the cause of much confusion in subsequent leasing and development.

Before the date of expiration of the Osage ten-year lease came around, the Cudahy company had resumed operations at the first well at Bartlesville, capped for five years because of Federal cancellation of the original Cudahy lease, but reopened when a lease on a small area surrounding the well was approved in 1902. The Cudahy company made its first shipment of oil from Bartlesville by tank car to the refinery at Neodesha, Kansas, in April, 1903.

James McClurg Guffey, originally known as "the king of the wildcatters," and his partner, Galey, had leased extensive acreage from individual Cherokees just prior to the Curtis Act. When a hitch occurred in the approval of the leases Guffey and Galey sold their Kansas oil interests to the Standard Oil Company and started operations in the Texas fields, but with an eye still upon the Cherokee and Osage country.

Theodore N. Barnsdall, commonly known as "T. N.,"

had already entered the picture. "T. N." was one of three sons of William S. Barnsdall, an English shoemaker, who had emigrated to the United States some fifty years earlier, and who had drilled the second oil well in Pennsylvania, following the Drake well of 1859. "T. N." had oil—and other things, including a bit of TNT—in his blood. Gargantuan were his exploits, alike in oil, in politics and in carousal. A huge figure of a man, rough and ready for anything, his achievements and colorful personality are a treasured memory of a thousand men still active in the industry.

A story of the Paul Bunyan type, written by Harry Botsford for the *Oil Trade Journal* some years ago, might suggest the business as revealed in the character and accomplishments of T. N. Barnsdall. To be sure, Barnsdall had nothing to do with this particular story, but it does suggest the extent and manner of his operations.

"Man, but this was a busy place! I was just a lad and I got a job working at the cook house of Clem Carew's outfit. We boarded something like 700 men and I'll never forget the way we handled them. We had stoves built of boiler plate over a hundred yards long, where we baked pancakes. We mixed our batter in a sixty-barrel tank by machinery and one man stood all mornin' handlin' a hose and playin' it on that great big griddle. Saved plenty of pourin', that hose did.

"I was just a kid then, and mighty active; so they tied a side of bacon on each of my boots an' my job was to keep skatin' from one end of that griddle to the other—regular automatic griddle greaser, if you see what I mean."

When the first ten-year Osage lease came up for renewal, Barnsdall had not quite reached that status, but he was on his way. He had become a director and heavy stockholder in the Indian Territory Illuminating Oil Company. He was reputed to be in the good graces of Standard Oil to the extent of several million borrowed dollars. If that is accurate, and there is considerable testimony to indicate it, no better evidence of Barnsdall's recognition as a very important figure in the oil business of the United States thirty years ago could be asked.

"T. N." knew how to behave in the offices of the mighty as well as among a group of cronies in an oil-field bunkhouse. He was one of the group which lobbied in Washington for months for the renewal of the Osage blanket lease. Sublessees who had struck oil on their land were the chief opponents of the larger company. The prize was countless millions. The battle brought in the names not only of Barnsdall, H. V. Foster, the Standard Oil Company and Guffey and Galey, but such familiar names as Senator Chauncey Depew of New York, Governor Higgins of New York and Senator Harry S. New of Indiana. Guffey and Galey had returned from Texas and started subleasing from bankrupt independents on the Cherokee lands near Bartlesville.

President Theodore Roosevelt and Secretary of the Interior Hitchcock had been entrusted with authority to settle the controversy and extend the leases on such terms as they might see fit to exact. The fact that authority had been given them through a rider on an appropriation bill passed in May, 1905, aroused considerable suspicion of the good faith of the entire proposition. Friends of the Indians and enemies of the larger oil companies insisted that it was pure, or impure, politics, made effective too hastily, without due regard to Indian rights.

At that time the certainty that vast oil pools underlay thousands of acres of Indian lands had been sufficiently demonstrated to establish a price of ten and twenty dollars, and more, per acre as a bonus for leases. In addition to such bonuses the lessees were to pay the lessors royalties running as high as one-sixth of the production. It was while comparatively small leases were being valued at that rate that the question of renewal of the ten-year blanket lease on the vast acreage of Osage lands came before the Department of the Interior.

In the Osage Nation alone some 680,000 acres had been sufficiently examined and prospected by geologists and drillers to convince such men as Barnsdall, Guffey and Foster, and such companies as Standard, I. T. I. O. and Prairie Oil and Gas that the leases must be renewed. At the minimum bonus of ten dollars an acre which was being offered for leases on small proved areas subject to the disadvantages of offset wells, a lease on the entire area would have cost $6,800,000. A great deal of effective lobbying could be done with a small fraction of that sum. It was done.

Senator Depew of New York, Senator New of Indiana and others with influence in Washington were interested. The hand of the Standard Oil Company, partly concealed in the corporate glove of the Prairie Oil and Gas Company, made itself felt. Half the original leasehold was set aside, but the 680,000 most desirable acres were again leased to the leading operators for a ten-year period without any bonus to the Indians and on a one-eighth royalty basis. In that situation Governor Haskell of Oklahoma wrote an indignant letter to President Theodore Roosevelt. It is an illuminating document, as the following extracts indicate:

"I am informed on what seems to be reliable information that several conferences were held between the Interior Department officials and yourself on the one hand, and Messrs. Barnsdall and Guffey of Pennsylvania on the other, to bring about your making of this lease, and that the direct representatives who conducted the negotiations with you for this lease were Senator Depew of New York, Governor Higgins of New York, and Harry S. New of Indiana, then prominently connected with the Republican National Committee.

"Mr. President, the presence of those five named gentlemen, including Senator Depew, who induced you, in the year of 1904, to order the granting of the pipe-line franchise in Indian territory to the Prairie Oil & Gas Company, should have been sufficient evidence that you were dealing directly with the Standard Oil Company, and therefore should have warned you that this vast property interest of these Osage Indians, of which you were in fact chief trustee, should have been the subject of the most careful consideration. And yet you did not even grant the request of the Osages that you give them an opportunity to offer evidence to you of the real value of their property. It is a low estimate, Mr. Roosevelt, to say that you should have obtained for them a one-sixth royalty of the production, and in addition to that at a very modest estimate at least $7,000,000 cash bonus. And then their leasing to actual operators in small tracts would have left the Standard Oil Company a handsome profit indeed, a fabulous sum, beyond the comprehension of ordinary minds."

That letter of Haskell's accomplished nothing more than the revival of indignation over a *fait accompli.* It may have had some bearing upon President Theodore Roosevelt's subsequent activities as a trust-buster, but that is mere guesswork. In the meantime, however, the Interior Department had promulgated a restriction of the size of leases on Indian lands outside the Osage reservation. Development of oil properties in the territory was proceeding with such a rush that applications for leases were smothering the Department. But that ruling merely prepared the way for more shenanigans.

A dispatch from Washington to the Muskogee *Times-Democrat* under date of May 1, 1906, reveals the righteous indignation of residents of the territory who felt that oil business possibilities and oil profits were being snatched from under their noses.

"The Interior Department has ruled that no oil person, natural or corporate, can hold more than 4,800 acres of land for gas and oil purposes," said the dispatch. "This applies only to the lands the title of which is still vested in the allottee, it being held that the department has no control over land held by freedmen and sold or land sold by Indians whose restrictions have been removed.

"The Prairie Oil & Gas Company and each of its principals therefore picked out 4,800 acres of the best land they could find, and still hold it. The next move was to establish sub-agencies or dummies with which to defeat the law . . . .

"Oil men who take up 4,800 acres of land must show that they have enough money to take care of that much land. The department seems to think that about $40,000 in cash is about right, and this much money has been deposited in a bank as either a guarantee of good faith or ability to lie. The latter seems to have been the case in most instances for the Standard has made it possible for ten-dollar-a-week clerks to make affidavit that they have $40,000 in bank. It has also made it possible for banks to make affidavit that the dough is there all right—at least long enough for the ink to dry on the affidavit. Armed with these credentials Standard Oil agents proceeded to take up $200 land at twenty-five cents an acre and fool the government.

"It was stated by an operator that at least 1,000 acres of leases worth at least $250,000 were taken by James F. Furlong, E. P. Whitcomb and others who were tools of the Standard, for which the allottees received fifteen cents an acre or less as 'bonus' money. George R. Smith, a former government employe, is said to have 'turned' many acres of valuable lands for from $5.50 to $25 per acre.

"T. N. Barnsdall is said to have violated the regulations . . . by having employes take leases in their

names for him. He is said to control 30,000 acres in
this way. He put up $40,000 for John C. Furlong,
E. T. Whitcomb and others. Barnsdall, Guffey and
Galey and others known to be connected with the
Standard Oil Company are campaigning to control
the whole output of oil and gas in Indian territory . . .
    "After gobbling up all good lands Standard Oil
lays other lines for control . . . Independents find it
extremely difficult to get their oil into the pipe lines.
They only get fifty-two cents a barrel for the same
kind of oil that brings $1.50 in Pennsylvania. The
small fry hasn't much chance unless it lets the Standard do the drilling.
    "Any operator who has likely looking leases can
get his wells drilled . . . Standard will always lend a
helping hand . . . They only take sixty per cent of
the proceeds; lessee thirty, and the Indian whatever
is left. If Mr. Operator doesn't like to do it this way
he can drill his own and let his oil go to waste. . . .
    "Not long ago there was a meeting of Independents
at Bartlesville. A delegation was sent to Washington
to set up a howl. Next an inspector was sent to Pittsburgh to look into matters. Barnsdall, Guffey and
Galey and some others were subjects of the investigation. Results are just leaking out by degrees . . ."

    In the meantime the Dawes Commission was continuing
its work under the authority of the Curtis Act of 1898.
Not until the discovery of the Glenn Pool near Tulsa did
the possibilities of oil have much influence on the value
of lands in the Indian territory outside of Oklahoma Territory. Most of the Indians of the Five Tribes sought allotments of sufficient size and fertility to provide them with
a living by farming. The more backward Indians made
no rush for their allotments. The more intelligent and
industrious obtained what appeared to be the choicest
acreage. When the more promising agricultural areas
had been allotted the less promising were assigned to

tribesmen, who, in some instances, did not even take the trouble to register their claims. Some, especially the older full-bloods who were still loyal to tribal traditions, actively opposed allotments under the new order. One protesting group of thirty or more under the leadership of Crazy Snake was jailed for resistance. More ironic jests of Fate were the eventual outcome of that procedure.

The most notorious example probably is that of the late Jackson Barnett. Barnett, marked down as an "incompetent" in the Dawes Commission census upon which all allotments were made, and provided with an artificial guardian, as were numerous others who attained less publicity, was arbitrarily assigned an acreage so barren and rocky that it would hardly support a jackrabbit. But when oil was discovered in that region the "incompetent" Indian's acreage proved to be one of the richest in the entire state. The fortune which came to him through the sale of a lease and resulting royalties gained nation-wide publicity because of the economic, legal and domestic ramifications which made it news for years.

Literally thousands of persons who have seen the senile old Indian gravely if impotently directing the streams of traffic on Wilshire Boulevard, Los Angeles, from the sidewalk corner of his expensive home, have smiled, and commented, and wondered. Hundreds of thousands have read newspaper stories of the domestic problems and resultant court actions evolving from that fortune. He is dead now, but the fortune and the legal battle over its management and distribution still continue. The incidents have produced a distorted popular impression of the whole subject of extraordinary wealth attained by Oklahoma Indians through the production of oil from their lands.

To explain the situation briefly and broadly, it should be understood that when the Dawes Commission allotted lands to individual Indians it gave them absolute title.

They could do what they pleased with their property, subject only to the restriction of legal guardianship imposed upon those enrolled as incompetent. They could farm, or lease, or sell their own acres as they saw fit, or their guardians could do it for them.

The result was, as in the case of Barnett, that those whose lands were underlaid with oil became wealthy as fast as oil development brought men and companies into their neighborhoods to buy leases, drill wells and produce fortunes on a royalty basis which usually meant one-eighth of the income to the lessor. That arrangement accounts generally for individual fortunes among members of the Five Tribes while other members of the same tribes remained in poverty. By an agreement reached in 1902 unassigned areas of Chickasaw and Choctaw lands were reserved to be sold for settlement, with the mineral rights reserved for the tribes.

The Osage tribe's rights were more or less conserved by a different arrangement. John Palmer, a Sioux Indian who had been adopted into the Osage tribe, is generally given credit for that. The Interior Department made a careful census of the Osage Nation in 1906 when the oil boom was just getting well under way, and enumerated 2,229 members of the tribe entitled to equal shares in the tribal lands and profits from those lands. To each of those individuals, regardless of age or condition, was assigned what was known as a headright. That was a 2,229th share in the mineral production of the tribal lands, to be collected and distributed under Federal authority. Every Osage Indian enumerated in that census, or his heirs, has been a rich Indian since that day.

In the next twenty years while the oil income waxed and waned the numbers of the tribe, the holders of headrights, steadily decreased. But each headright was a valid title to a specific equal share in the whole tribe's total income from oil. In that period some members of the tribe acquired by inheritance as many as eight headrights, which brought some incomes above $100,000 annually.

Those incapable of attending to their own business because of youth, senility or other reasons were placed under the legal care of guardians assigned by the Federal government. Originally there were 625 of those so-called incompetents. What some of the guardians did to their wards was a scandal and a crime, but as time went on that situation, generally, improved.

Broadly speaking, all the Osages were cared for and their prosperity assured when the original Osage ten-year lease was in part renewed. Other tribes were not so well protected. That was the situation while the question of statehood was still agitating Oklahoma Territory and the adjoining Indian territory, while the last frontier still retained numerous characteristics of a frontier. It was soon to be altered.

CHAPTER XIII

Riches End the Frontier

Plans for the advancement of the Territory of Oklahoma to the status of the State of Oklahoma had been under way almost continuously since the establishment of the Territory. The first statehood bill had been presented to Congress and rejected as early as January, 1892.

The Five Civilized Tribes, exerting the strongest influence in Indian territory, opposed all bills which would include their more or less independent "nations" in the proposed state, and thereby place them under the political domination of the whites who had been admitted to their ceded lands in the West. The battle was fought in Congress and out, for years. Oklahoma Territory in the course of time temporarily abandoned its hope of including the Indian country in the proposed state, and sought admission as a state with the same boundaries as the Territory. The Indians, on the other hand, went so far as to draft a constitution for a separate state to be called Sequoyah. Separate statehood and single statehood parties carried on the battle.

Oklahoma Territory had a dominating white population, and the advantage of official representation in Washington. But it was still an agricultural region. What wealth it enjoyed was largely in the hands of a comparatively few prosperous cattlemen. The Indian country had a very small proportion of white residents, and most of those were intermarried with the Indians. But oil was making some of the Indians wealthy and attracting the attention of powerful corporations. The resultant pressure for single statehood brought about passage of an

159

Enabling Act, signed by President Theodore Roosevelt in June, 1906, preparatory to adoption of a constitution and organization of a state government. Those details accomplished, and Charles N. Haskell elected governor, President Theodore Roosevelt, on November 16, 1907, proclaimed Oklahoma the forty-sixth state of the Union.

The frontier era was finished, except in so far as development of isolated regions was concerned. Oil discoveries and consequent riches had exerted pressure and prepared the way for development along the familiar lines of the modernization of all the other western state areas.

The night of declaration of statehood for Oklahoma was one unforgettable by those who took part in the celebration, and that included almost every white person and many Indians throughout the state. A monument in a cemetery at Bartlesville not only puts the date into stone but reveals that the character of the frontier could not be so easily or so abruptly altered as its erstwhile system of government. The granite bears a simple inscription:

Earnest Lewis
Killed by Fred Keeler
Nov. 16, 1907

It is an unconventional gravestone, to say the least, and worthy of the last frontier. It is not very difficult to find men in Bartlesville who remember the incidents of that night. It all happened more than thirty years ago, but it left an impression graven upon the minds of the old-timers as deeply as it is graven upon the granite.

Earnest Lewis, according to my informants, was a man of frequently fictionized western type. In the course of his business operations he had been annoyed by certain U. S. deputy marshals. When he established himself in Bartlesville he warned two deputies, Fred Keeler and George Williams, that if they entered his place of business they would do so at their peril. In the frontier days that was generally accepted as fair and sportsmanlike practice.

But everything was wide open in Bartlesville on the night of statehood, as it was throughout the new state. Men and women paraded the streets, singing and shouting their delight. Everyone was joyous. All enmities appeared to have been forgotten. And so, in the course of the rounds of revelry Keeler and Williams happened into the Lewis establishment, filled with the spirit of good fellowship. But Earnest Lewis didn't know that. He knew only that he had warned them according to the code of the frontier to stay away.

He promptly unlimbered a six-gun and started to shoot. Williams dropped, mortally wounded. The gaiety, good fellowship and hilarity of that crowd went out like a light, and the crowd did its best to follow. But Keeler stood fast, pulled his own gun and, shooting over the body of his fallen comrade, dropped Lewis in his tracks. Bartlesville's celebration of statehood for Oklahoma came to an abrupt and dramatic end. The frontier spirit had not been entirely eliminated by the mere formality of turning the frontier into a state. Lewis's widow was a woman of spirit, as women of the frontier should be. She revealed the frontier spirit once more in the granite stone upon her husband's grave: "Killed by Fred Keeler." It scandalized the more conventional residents of Bartlesville, but left them powerless to change the record.

Oklahoma, an Indian word for "the land of the red man," was a state. The red men were still an important influence in its eastern portion where oil was producing an economic and social revolution. The Indians were not always the honest, innocent children of Nature which sentimental writing has suggested. Some were as devious and unscrupulous as the whites who sought to exploit them, although usually with more limited imagination.

For example, when the first land office was opened in the

Chickasaw Nation for enrollment of Indians for allotments under the Dawes Commission, the official in charge cautioned employees to be patient with the "ignorant" Indians. When the doors were thrown open a long line of men and women of all ages and conditions began to press toward the counter. The official, noticing one feeble old woman tottering in the line, besought the man who stood ahead of her to give her his place. The man refused. The official thereupon led the old woman from the line to a rear door, took her in and made out her registration certificate. Before he had finished a second squaw entered from the rear and asked similar service so that she could take the old woman home. Another and another found excuse for precedence bearing on the same case. Eighteen had thus gotten ahead of the line before the official realized that he was being tricked by the "ignorant" Indians. Hasty investigation revealed that two Indians outside were rounding up the squaws, providing each with a suitable story and moving them through the back door.

That was a petty fraud, interesting only as it indicated the fact that the Indians were as willing and eager to trick the white man as the white man had ever been to trick the Indian. But most of the latter lacked the imagination to carry their trickery to more profitable lengths. At times one appeared with a more effective plan of defrauding the white brother.

In the early days when independent oil operators and lease buyers were most active, some of the shrewder and less scrupulous Indians sold leases on their allotments for cash to every buyer who applied. It is recorded that one old Cherokee signed six leases, for cash each time, on the same allotment. That was made possible in part by the fact that Secretary Hitchcock's arbitrary restrictions upon actual drilling prevented each lessee from starting a well immediately, and none of the buyers was aware that the privilege of drilling had been sold. Then all the

An Oklahoma country road and its mud-made traffic jam on the way to a new oil field.

buyers rushed in with their drilling rigs only to find that the "ignorant" Indian had taken the cash, in effect multiplying his legitimate bonus by six, and left them to scramble over what they had purchased.

Many of the Indians cashed in their allotments or their lease bonuses as quickly as law and circumstances permitted. It was the usual practice of relatives and friends to descend immediately upon the beneficiary of momentary affluence and make themselves at home with feast and frolic until all the money was squandered.

An exception to that sort of potlatch hospitality found its way into the columns of the Muskogee *Times-Democrat* of December 27, 1906. A Creek known only as Charley came into sudden comparative riches. But when his friends arrived for what they expected to be an extended and informal house-party, Charley had vanished. A week or two later he reappeared, looking somewhat the worse for wear, and explained sadly that he had spent all his money drinking and gambling in Oklahoma City. The relatives mourned, and explained regretfully to friends:

"Too bad! Charley heap damn fool. Went Oklahoma City. Got drunk. Gambled all his money. Damn fool. Indian damn fool anyway."

But subsequent developments revealed that before his disappearance Charley had carefully deposited most of his money in a Muskogee bank where it remained to be used as needed in farming land which he still controlled. All the Indians were not damn fools.

Oil prospectors and lease buyers did their best to make fools of them. Frequently it did not work. Another item from the Muskogee *Times-Democrat* makes the point.

"Now you see," said one high-pressure lease buyer to an Indian whom he considered a prospect, "in thirty days (a lie) there will be oil wells dotted all over your land."

"Ugh!" was Lo's response.

"And your children can go to Yale."

"Ugh!"

"And your wife can have a new red silk dress every day."

"Ugh!"

"You see, it's like this: Oil's eight dollars a barrel (a lie) and out of every one hundred barrels you get ten—understand?"

"Ugh! Where hell get um barrel?"

Adventurous young women pursued a more effective course. When the news of quick and easy fortunes among the Indians in the neighborhood of Muskogee began to appear in newspapers of neighboring states a Miss Edith Davis of Springfield, Missouri, appeared at the Hotel Hoffman in Muskogee and applied for a job as waitress. Local business was excellent. The town was in its first flush of prosperity through allotments and oil leases. The girl went to work. On Sunday morning she met Louis McGibbon, a wealthy and influential member of the Creek tribe. On Monday morning they were engaged. On Tuesday evening they were married. The story received widespread publicity. A flood of potential waitresses of all colors, conditions and creeds poured in on Manager Hoffman of the hotel. Some were hired and succeeded in their schemes to get rich by marrying wealthy Indians. Others were frank but not so fortunate.

An incident is recorded of a beaming Swedish girl who announced with disarming candor: "Ay tank Ay want a yob. Ay come all the way from Meeneapolis, an' Ay tank Ay lak a Injun maybe so lak Meesouri girl."

Although the majority of the Indians lacked the experience and education necessary to take full advantage of their opportunities, shrewd ones overlooked no chances to obtain as many allotments as possible for the various members of their families. Altogether some two hundred thousand claims were registered by the Dawes Commission and some ninety thousand were allowed. The difference represents the number of frauds attempted by the Indians or by whites operating in their names. Investigation revealed some interesting situations.

A thirteen-year-old girl led before the Cherokee enrollment division of the Dawes Commission was asked who gave her the name which was offered for registration. "That woman," said the child, pointing to the squaw who accompanied her. The woman claimed to be her mother, and managed to prove that the child was entitled to enrollment as a half-blood Cherokee.

But cross-examination revealed the fact that the girl had been under the care of a mulatto woman in Kansas to whom she was given when she was only two weeks old. Only when the potential value of her rights as a Cherokee had been impressed upon her mother had the child been taken from her foster mother through a charge of kidnaping and habeas corpus action in a Kansas court. The foster mother, she said, had always treated her kindly.

"I'll run away from my mother as soon as I can escape," she told the commissioner.

But her right to an enrollment was clear, and the registration was made. Whether child or Indian mother eventually profited is not in the record.

In the meantime business was booming through most of eastern Oklahoma. The Glenn Pool south of Tulsa, the pools around Bartlesville and in the eastern Osage country, and the Muskogee area were the chief sources of excitement, riches, and general activity. Under the influence of oil the cow-town of Tulsa had grown from a population of 1,390 at the turn of the century to an oil-town of five times that number on the birthday of Oklahoma's statehood. Muskogee in the same seven years had tripled in population and boasted some 13,000 residents.

Muskogee had been the seat of Federal administration of Indian affairs in the territory for thirty years. With the expansion of activities of the Dawes Commission its chief source of income had been a Federal payroll. Even oil could not entirely overcome that handicap. It could add to wealth, but it could not overcome the inertia of a large force of government payrollers. Tulsa, on the other hand, was on its own. It could look to no paternal government for support. It must work or starve. It must seize and improve its opportunities or see those opportunities go elsewhere. Its spirit was the spirit of pioneers, of men and women able to withstand privation while they sought and grasped every opportunity to improve their condition.

Their first great opportunity appeared in the overflow of the initial rush of promoters, lease-seekers, well-drillers, supply men, speculators and so forth into the neighboring Red Fork district as soon as Kansas City and other newspapers announced the gushing of the Sue A. Bland well. Red Fork could not accommodate the crowds. Tulsa did its best to serve.

According to J. M. Hall, a partner in the first store in Tulsa, the town had a hotel in the first year of its settlement, 1882. According to Lon Stansbery, the first hotel was one sponsored by W. M. Robinson of Kansas City, who remodeled a livery stable at First and Boston streets. "It took six months to get the horseflies out so the citizens could enjoy a real hotel," says Stansbery. With practically no accommodations at Red Fork, that hotel did a profitable business in the first oil rush. Few of the residents of the village, then numbering approximately 2,500 persons, including Indians, children and cowhands sobering up between shipments of cattle, were too proud to help accommodate the overflow. The few local stores transacted more business in a month than they had transacted in any previous year. Sales of Peruna and Jamaica ginger were enormous. The villagers handled more money than they had ever imagined they would see, and they liked it.

When the Glenn Pool came in they took prompt and effective steps to see that Tulsa became its metropolis, its source of supplies, its point of accommodation for office space, banking service and living quarters for all those not physically employed at the scene of drilling and production. So early Tulsa recognized its possibilities and began to build toward its proud, and accurate, boast of being the "Oil Capital of the World."

When Sapulpa, nearer railroad point to the Glenn Pool, made its bid for supremacy by erecting a huge sign at the station announcing what a fine town it was, Tulsa countered with the construction of the Brady Hotel, widely proclaimed as spacious, not to say luxurious. Such accommodations proved more effective in their appeal to oil promoters than was Sapulpa's painted sign claiming advantages which could neither be slept in, eaten nor used for office necessities.

In the year of Oklahoma's proud attainment of statehood the Glenn Pool attained a peak production of 117,440 barrels of oil per day. Numerous smaller fields had been proved around Bartlesville, in the Osage lands, and around Muskogee. Total oil production had advanced in four years from 1,367,000 barrels valued at $1,326,000 to 43,524,000 barrels valued at $17,513,000.

The Prairie Oil and Gas Company, child of Standard Oil, chief purchaser of crude oil in Oklahoma, completed a pipe line to the refineries at Whiting, Indiana, 600 miles away. The Texas Pipe Line Company and the Gulf Pipe Line Company completed outlets to refineries and shipping points on the gulf. Everything seemed high, wide and handsome. And then came trouble.

A tale of Oklahoma's most famous "Coal-Oil Johnny," told in detail in the Kansas City *Star* of December 21, 1913, may help to reveal the picture. Billy Roesser was one of the first young men to cash in on the tremendous riches of the Glenn Pool. Billy grabbed leases right and left before greater, more stable, and wealthier men and companies reached the spot. When the scramble of the industry's leaders started with the proving of the pool

Roesser quickly sold eighty acres of his holdings for $350,-
000 in cash, and retained leases appraised at $1,000,000
more.

That was a lot of money for a young man of thirty-two.
It had come so swiftly and easily that the young man knew
it could be duplicated whenever he cared to apply his
talents. The finest house in Tulsa was not too good for
such a captain of finance. The finest house at the time was
one owned by George Bayne. Billy had never seen it, but he
had heard of it. He promptly offered Bayne $35,000 for
the house and furniture, unsight and unseen. The deal
was closed.

When the new-rich young man went out to see what he
had bought, a companion told him that the mahogany
furniture was only veneered. Nothing like that would do
for an oil king. He told his friends to send out trucks
and haul the stuff away. It was theirs for the taking. Then
he engaged decorators and furnishers from St. Louis
to do their best. When the question of pictures suitable
to the owner's high estate arose, he sent for an art dealer
from New York and ordered $16,000 worth of "real genu-
ine oil paintings."

But the garden area was wide and bare. That also was
simple for a man of Billy Roesser's talents. He had only
to order a trainload, fifteen cars, of bluegrass sod from
Neosho, Missouri, have it laid on nearly two acres of bare
ground, pay out $7,000, and the problem of a lawn was
settled almost over night. A landscape gardner from St.
Louis put in $8,500 worth of shrubs, trees and flowers.

An establishment of that kind required a housewarming in keeping with its elegance. There was no question
of finding plenty of friends to attend the party, but there
was a doubt that anyone in Tulsa thirty years ago was
competent to arrange a party adequate to the occasion.
Billy journeyed to Kansas City, engaged the entire cater-

ing force of the Morton establishment, hired the Carl Busch orchestra and put on a housewarming that cost him $5,000. Needless to say, it still remains a highlight in the memories of those who still survive.

Automobiles were just coming into popularity. When Roesser was advised by a Kansas City agent that he had a very superior machine, painted and upholstered in a style suitable to an oil king, at a price of only $6,500, Billy wired to send it on to Tulsa by express. A little later, at an automobile show in Kansas City, he paid $14,000 for six cars at once.

Driving one of them to examine an oil property in which he was interested he bogged the machine in the mud. There was not a paved highway in Oklahoma in those days. With considerable difficulty the young oil king got the machine back to town. He was standing on a plank sidewalk looking it over when a friend came up.

"What'll you give me for her, Tim?"

"Two hundred dollars," said Tim.

"She's yours," said Billy.

Still Billy Roesser retained his interest in motor cars. When he heard of the first motor races planned for the Tulsa track he wired an Indianapolis firm to ship him a racing car at once. It arrived just in time to be entered in the race. When an automobile racing meet was advertised to be held at Crown Point, Indiana, Roesser invited three friends to go with him. "I just want to show you a good time. It's my treat," he said. The little outing extended from Tulsa to Chicago, Crown Point, Indianapolis, Columbus, Cleveland and back to Tulsa at a cost of $2,000.

Billy was no piker. When the church of which he was a member asked for contributions he gave $5,000. Any of his old friends of the days preceding his attainment of wealth and glory could touch him for a hundred or a thousand. He scattered largess alike among friends, relatives and indigents. He grubstaked scores of oil prospectors and lease-buyers.

When rumors of the discovery of oil in Bond, Clinton and Montgomery counties, Illinois, reached his ears Roesser was beginning to need money. It was surprising how fast a million and a quarter could disappear. But Billy knew the game. He organized a syndicate and leased 110,-000 acres in the new field. Every well sunk on that area was a duster. Undisturbed, he leased 8,000 acres near the Glenn Pool, where his first success had gushed to his head and his pocketbook. It cost $130,000 to sink nothing but dry holes there.

Well, that was the oil business. From rags to riches in a single year. From riches to rags in another five years. Billy Roesser had something besides an inflated ego. He had the spirit of his time. He had done it once; he could do it again. He sold the big house, the imported lawn, the "genuine oil paintings" and most of the furniture, and rented a modest cottage where he established his wife and two children. With that small capital he went out in deadly earnest to recruit his fortunes.

Not more than a year or so later on a winter night he climbed down from a day coach in Tulsa and walked briskly home, cheerfully fingering his total capital of forty dollars in one pocket. When his wife wept at his latest news Billy grinned and patted her shoulder. "That's as much as I had five years ago, isn't it?" he said. "There's nowhere to go but up."

And twenty-four hours later he started a second ascent. An oil prospector whom he had befriended met him by chance on the street late at night and told him that a new well had just come in on a certain tract east of the Glenn Pool. Instead of going home Roesser went to a garage, hired a car, and drove away into the winter darkness. At daylight he stood beside the new well. Within an hour he had leased a nearby tract from a farmer. Before noon he had sold it to a speculator at a profit of $600.

With that money he hurried to the Cleveland field, just coming into production, and leased a piece of ground. He still had friends who had faith in his luck, or who were ashamed to turn him down. He borrowed enough from them to sink a well. He really did know something of rock and sand formations. When the sand raised by the drill indicated to him that there was oil beneath, he promptly shut down operations and went out to lease more ground. He had no money, but he had a reputation. Several property owners agreed to put leases in escrow and allow him a few days to pay. Then he renewed the drilling of his own well. It came in with a rush and continued to flow at the rate of 200 barrels a day. He sold his leases for $20,000.

That money promptly went into a lease on 120 acres south of Cleveland. Friends who were regaining respect for Roesser's oil judgment took a three-fourths interest in return for enough cash to sink five wells. Roesser won $50,000 cash from that gamble. A similar venture near the Cushing field netted him $20,000.

With $90,000 of his own available Roesser made the plunge, with the lease on twenty thousand acres near Meramec. "That's going to clean me up fifty millions," he said.

"Suppose you lose."

"Well, suppose I do. In this oil business a man's got to be game. Risk it all for a big stake."

"If you strike it rich again will you be as free with your money as you were before?"

"Sure. What good is money except to have a good time with and help your friends with?"

And so, a noteworthy and entertaining example of oil philosophy, oil success, and oil failure, Billy Roesser calmly watched his final $90,000 gamble swept from the board. Probably he figured he had lost $50,000,000. He never came back.

In contrast, another oil man of humble origin, similar to Billy Roesser only in his willingness to back his judgment and his activities to the limit of his resources and to spend his profits in more spectacular and more permanent manner, actually did make, spend and lose several fortunes publicized at from $12,000,000 to $50,000,000 each.

Because of its spectacular ups and downs and its present promise of final stability the career of Joshua S. Cosden sheds a light on the possibilities of the oil business quite different from that of most tales of oil-made fortunes. Old-timers of Tulsa who have seen their city built and who have helped to build it to greatness with the profits of oil are still conscious that "Josh" Cosden gave to their city and to all of eastern Oklahoma more than the world's largest independent oil refinery, more than the substantial Cosden Building, more than the development of wells, pipelines and similar sources of income.

Both in his successes and his failures he brought to Tulsa and to the Oklahoma oil fields and to the various branches of the Oklahoma oil business an invaluable energy, imagination, daring, optimism and publicity, when all those qualities were most important to the city's development.

Joshua Cosden found his first job, at the age of twenty, as a reporter on the Philadelphia *Public Ledger,* not far from his home town of Baltimore, Maryland. He still recalls that experience with satisfaction, but it was not the experience for which he was destined, either by Fate or by his own driving energy and consuming ambition. He gave it up to enter the real estate business in Baltimore.

There he made some money and more friends, and perhaps learned some of the rudiments of the science, art or business of promotion. He was alert for opportunity, and when an acquaintance suggested a new theory of oil refining and gasoline extraction he recognized Opportunity with a capital "O." Automobiles had started to establish themselves as a factor in the life of America. Improved lubricants, and better and cheaper gasoline, appeared to be a growing necessity, of unlimited possibility. The Oklahoma oil fields, especially within the Osage reservation, were beginning to attract wide attention.

Cosden put the two ideas together, interested Baltimore friends who had confidence in his judgment, ability and energy, traveled out to the wilds of the Osage country, and built at Bigheart what he says was the first refinery in the country to crack gasoline by a continuous distillation process. The immediate trouble was that he did not have much oil to run through his refinery. His sources of supply were wells some three or four miles from his plant, capable of producing a few barrels of crude per day. There was no pipe line, and not even a road between well and refinery.

The crude was transported in a second-hand tank wagon, leaking copiously at every strained seam as it jolted and labored over the rocky terrain, with Cosden plodding behind, holding a bucket under the worst leaks and pouring the recaptured oil back into the decrepit tank from time to time. He considered himself fortunate if he could get two and one-half barrels of crude into the refinery on each laborious trip. Eighty barrels of oil to the refinery in one month was a fair run.

The refinery building was blown away once by a cyclone. The cabin in which Cosden lived was wrecked at the same time, and for two weeks he made his home in a hole in the ground which served as a cyclone cellar. The plant burned to the ground twice. But Cosden persisted. He even found a little relaxation and social diversion, playing pitch with Red Eagle and his wife Rosie Red Eagle at a penny a game.

"That was amusing," he says, "but after a while I began

to wonder why I always lost. Finally I decided that it might be because Rosie and Red Eagle were speaking Osage all through the games, and I couldn't understand what they were saying. So I induced the local livery stable man, who was also justice of the peace and understood Osage, to go with me. After that I frequently won."

Years later, when Mr. Cosden was a director of the Midland Valley Railroad on his way in a private car to inspect the Burbank oil fields, his train stopped at Bigheart. Memories of the old hard days crowded in upon him. He sent word to his old friend the liveryman and justice of the peace, invited him into the private car, and the train was held up for half an hour or more while the two one-time pitch-players roared with laughter over reminiscences of those card games in an Indian cabin.

Long prior to that incident, of course, Cosden's supply of crude, his transportation facilities, his output of gasoline, kerosene and refined oil had greatly improved, and his financial situation and ambition had improved with them. Through the same early years the oil fields were extending in area and production. In 1912 Cosden started his first large refinery in West Tulsa, financing it in part with his own money and in part with Baltimore money.

The refinery continued to expand through the next four years. Then Cosden decided that the big money was to be found in Wall Street rather than in Baltimore. So in 1916 he decided to go into the production end of the business. He formed a company for that purpose and took his proposition to one of the largest investment banking houses in New York.

"The bankers put up $12,000,000 on the proposition which I made them at that time," he says. "I walked into their offices at two-thirty in the afternoon and walked out at four-fifteen with the money assured."

Much of that money was used for the purchase of producing wells, mostly in the Cushing field. That was Mr.

Cosden's entry into the producing end of the business. Prior to that he had been solely a refiner. In order to take care of his newly acquired production it was necessary, after war business had stimulated demand in 1916, to expand the refinery to the extent of installation of 130 additional stills at a cost of approximately $5,000,000. One hundred of those new stills were the first cracking stills installed by anyone other than Standard Oil companies. Because of patents and licensing restrictions placed by Standard Oil on the use of such stills, Cosden developed his own patents for the cracking of petroleum products. The completed refinery was the largest independent in the world. Approximately $28,000,000 was invested in the plant. By the addition of the cracking stills Cosden was in a position to improve his output and overcome the handicaps which beset others. He built a lubricating plant at West Tulsa that was declared to be the finest in the world, and expanded his fleet of tank cars to 2,000 units.

Those years of development and expansion in Tulsa put Cosden definitely into the big money. He not only acquired it and used it to expand his business interests, but spent it lavishly in personal and social activities. The home which he built in Tulsa was one of the residential showplaces of the state, in some respects pointing the way even to the homes of movie stars of a later decade in Southern California. A feature then almost unique in home building was an indoor swimming pool. In the garden were two tennis courts for which the clay was said to have been imported from France at a cost of $10,000.

But Cosden's activities outside his success in oil were not limited to social and personal triumphs. Tulsa had been good to him; he would be good to Tulsa, and at the same time diversify his investments. He proved it with a modern sixteen-story office building at Fourth Street and Boston Avenue, on the site of the first mission church and school of Tulsa. The Cosden Building is still Josh Cosden's monument in Tulsa.

It proved to be something more than an office building. Its roof was the site of what was widely acclaimed to be the first residential penthouse west of the Mississippi River. It was occupied by Cosden after he gave up the original residence which had excited the admiration of Tulsa's wealthiest citizens.

Josh Cosden's name had acquired magic in the booming Tulsa of the war-stimulated oil business. Tulsa was proud of him and his accomplishments. It profited indirectly from the attendant publicity which brought not only Cosden's name but Tulsa's name and mention of Tulsa's wealth of opportunity more and more frequently into New York newspapers, as Cosden made more and more frequent trips to New York and became a more and more familiar figure in Wall Street.

He was a multimillionaire. He had both social charm and social position. He was a plunger and a spender. All that assured popularity in New York. He accumulated more millions—so many that even the conservative New York *Times* once reported an estimate of $50,000,000.

He purchased a magnificent estate at Port Washington, Long Island, and acquired a winter home at Palm Beach, Florida, suitable to his fortune and his reputation as a host. Yachts, private cars and other appurtenances of wealth and facilities for entertainment were in keeping.

In that situation Mr. and Mrs. Cosden achieved their greatest social triumph when they entertained the Prince of Wales and Lord and Lady Mountbatten as house guests at their Port Washington estate. There is no definite testimony available as to precisely how deeply that incident impressed the New Yorkers, but the impression upon residents of Tulsa was universal, deep and permanent. Of a score or more persons who told me of Josh Cosden's part in the building of Tulsa, not one, I believe, failed to comment on the fact that Mr. and Mrs. Cosden entertained the Prince of Wales. Mr. Cosden now deprecates that publicity.

In the less snobbish story of material accomplishment Josh Cosden's career is more dramatic. When widespread overproduction of oil in the American market seriously impaired oil profits Cosden decided that he must devote more time to his personal interests. In doing that he gained control of the Lago Transport Company in Venezuela, later a producer of oil at the rate of 130,000 barrels a day for years. He also controlled the Creole Oil Company, another Venezuela concern. Both great concerns eventually came into the control of the Standard Oil Company of New Jersey.

But in the meantime Cosden's Venezuelan and Wall Street interests precipitated a situation in which he was forced to choose between them and the presidency of Cosden & Company, with the demand which the latter responsibility made upon his time. He preferred to give up the presidency.

In the same period he had been buying heavily of the stock of the Cosden company in an effort to hold it up on a falling market. That was a matter of pride. Eventually he found himself so over-loaded that he could no longer control the situation.

At last he awakened one morning to find himself broke. In those days to have less than a million dollars was to be broke. But he had terrifically heavy obligations, and comparatively little cash.

Both the Long Island and the Palm Beach estates were sold, at a noteworthy profit over their original cost. Most of the other tangible evidence of wealth and social accomplishments was sold. Tulsa was shocked. But Tulsa still had faith—and justification.

Less than two years later the New York *Times* devoted an entire column of its space to a story of Josh Cosden's spectacular recovery of a fortune of $15,000,000 in oil and the development of a $25,000,000 company again under the name of Cosden. That was in August of 1929. And fifteen months later that fortune followed the earlier fortune, into the limbo of receivership.

"And Joshua Cosden probably will roll up his sleeves, come west again and start out on his third fortune," said the Oklahoma City *Times* of November 10, 1930. It would be hard to find more impressive testimony to Oklahoma's faith in its most spectacular tycoon, especially in view of the fact that after the earlier crash Cosden had transferred his interests and activities to the Texas fields.

Further justification of that faith is revealed in a news dispatch printed two and one-half years later in the *Daily Oklahoman*: "Joshua S. Cosden, who made and lost an oil fortune of $50,000,000, is attempting another comeback. The colorful oil pioneer has regained control of the Cosden Oil Company. The company's properties, all in Texas, were purchased by Cosden recently at a receiver's sale at Big Spring, Texas. Cosden, whose bid was $501,000 represented a reorganization committee seeking to lift the company from the receivership into which it sank in 1930. . . ."

In December, 1937, four and one-half years after that dispatch was printed, I had the good fortune to meet and talk with Mr. Cosden in Palm Springs, California. According to the information which I gathered in the course of our conversations, the best of those predictions have all been justified.

The Cosden Petroleum Company, organized in May, 1937, with both stock control and its presidency in the hands of Mr. Cosden, has not only reorganized and consolidated and improved the assets and markets of the once bankrupt company, but it is operating on what Mr. Cosden firmly believes to be the most substantial if not the only truly substantial basis of permanent success—

internal rather than external development. That, he insists, has been the basis of the success of Standard Oil, as it has been of the Ford Motor Company, and will continue to be the guiding policy of the Cosden Petroleum Company.

"It is the only substantial way to success with peace and tranquility," he told me, "and peace and tranquility are what I want."

Evidently some of the fire has burned out in Josh Cosden in thirty years of a life as intense as any man ever lived in the oil business or any other. It was a tremendous fire. Let him summarize it as he summarized it for me.

"When a man plays for high stakes every day of his life for years, when he races horses, fights economic and political hazards, and lives constantly up to the limits of his energies and his physical, intellectual and material resources, it burns him up. I have lived that way, and I have enjoyed it. But I have learned that I want something else. I want peace and tranquility. I am interested in a great many things—in world affairs, in a variety of things both intellectual and material.

"I have had a great variety of experiences. I know many leading men throughout the world, and some of them in a friendly way. I have had great times. I owned one of the finest racing stables and breeding farms in the United States in the early 'twenties. My horse, Paul Jones, won the Derby in 1921. My horses have won many of the great stake races. I quit all that to devote myself to business. You can't breed and race horses and attend to business at the same time.

"I found on the Oklahoma frontier in the early days of oil the greatest lure and the finest human and social drama in the history of America. The lure of oil is the most romantic business in history. It is a business on the highest plane because in finding and producing oil we are taking from the earth a hidden natural resource of

great value to humanity. We are hurting no one, and are
helping many.

"In that work I have made and lost several fortunes.
I have lived, I believe, almost to the top of human capacity to live. In doing so I have learned many lessons.
Perhaps the most valuable and satisfying is that there are
many ways to live. One of them is in peace and quiet.
That is the one I am enjoying now, and intend to continue
enjoying, both in my continued business activity and in
my recreation."

CHAPTER XIV

MONUMENTS TO THE OIL INDUSTRY

MANY men whose names and accomplishments are still famous and influential in the colossal oil business of the United States had obtained their start in the eastern fields of Oklahoma before the territory became a state. Most widely known of those still living is Harry F. Sinclair. Among pioneers of the industry who first established themselves in Indian territory and who are still active in Oklahoma, perhaps the names and accomplishments of Frank Phillips and H. V. Foster lead all the rest. Among the powerful pioneers now gone but not forgotten were T. N. Barnsdall and Charles Page. Not a pioneer, but a modern leader, is William G. ("Bill") Skelly.

As in the contemporaneous history of the automobile industry which helped to make possible the extent of the oil business, the names most familiar to Americans are the names which have been most widely emblazoned on filling stations, refineries and tank cars. Sinclair leads all those. Phillips and Skelly are close behind in the Middle West. Names such as J. E. Crosbie, R. M. McFarlin, James A. Chapman, Tom Slick, Wirt Franklin, Lew Wentz, H. F. Wilcox, Frank Buttram, Dan J. Moran, Charles Colcord, E. W. Marland and many others have been more or less overshadowed outside of Oklahoma itself, by the names of the companies which they developed.

Some of those men are still highly important figures in the industrial, political and social life of the country, but many names have taken a place secondary to their companies, just as the name of Henry Leland, for example, has been superseded by the name of his Cadillac automobile. At least they have been superseded in the consciousness of a hundred million Americans whose present manner of life is largely dependent upon the supply of gasoline and oil for motor power and transport. In Oklahoma, where some of the leaders still make their homes and maintain offices, a mere mention of those names brings forth a flood of stories of accomplishment.

Before we go on to consideration of the most widely publicized figure of them all—Harry F. Sinclair—this might be a proper place to sketch the accomplishments of one man whose name was never emblazoned upon a filling station but who has left a monument of which any human being might well be proud. The name is Charles Page. The monument is the Sand Springs Home.

There is, I believe, nothing to compare to it in the world. Many hundreds of men, women and children owe virtually all that they have to Charles Page's wise, shrewd, far-seeing sense of practical philanthropy. It is, financially, a product of Oklahoma's oil. It is also a product of a superb imaginative intelligence and philanthropic instinct fostered through a youth of self-dependence and a mature life of business success.

Charles Page was born in the little town of Stevens Point, Wisconsin, in 1861. His first job was that of a telegraph messenger boy. Soon he was a telegraph operator, then a miner, then a timber cruiser. At the age of twenty-one he became chief of police of Ashland, Wisconsin. Later he was an operative of the then famous Pinkerton Detective Agency. In that capacity on one occasion he trailed a murderer halfway across the continent, lived with him for several weeks, confronted him with the charge, obtained a full confession, and returned him to Chicago for trial. Still later he entered the colonization service of the Northern Pacific Railroad in the Northwest, and there became interested in mining.

That was his background when the discovery of oil at Red Fork brought him to Oklahoma. It was a background of practical experience which had made him in middle life a hard-headed, hard-handed, intensely practical man without rendering him callous to the problems and tragedies of the human beings around him. He could still remember the days of his youth when a sandwich or a slice of cake handed out by some housewife of Stevens Point to whom he was delivering a telegram had seemed a tremendous kindness. In fact he remembered those days and those kindnesses so well that when his oil operations in Oklahoma made him rich he provided annuities for three old women who had proved themselves friends in his childhood. One of those annuities amounting to $150 a month was still being paid from the Page endowment years after his death. The others were paid regularly up to the death of the beneficiaries.

Charles Page had the heart, though not the fortune, for that sort of philanthropy when he began oil operations on a small scale. His first two wells were dry. But he persisted. When the Glenn Pool brought in its first spectacular wealth, Page was one of the first independent oil men to profit. From that time on he was rated a millionaire. Businessmen, and especially oil men, also rated him as one of the shrewdest and hardest traders in the industry. His long and finally successful battle for the famous Tommy Atkins lease, to be told later in these pages, is an example of such practical accomplishments. But the Salvation Army of Tulsa was quick to learn that he had a heart.

J. Burr Gibbons, now a leading advertising man of Tulsa, was then a cub reporter on a Tulsa newspaper. Among the news sources which he cultivated was a young Captain Breeding of the Salvation Army. Captain Breeding frequently was good for a story about some accident in the oil fields which had been brought to his attention

through an appeal for help. Some family had been left
destitute by the death or injury of its breadwinner. There
were in and around Tulsa numerous families of oil workers who had been brought from Pennsylvania or elsewhere
into the new oil fields. There was no employer's liability
in those days. Families of workers who had been incapaci-
tated had to be supported by charity or sent back by charity to their erstwhile homes in other states.

When Captain Breeding revealed these pitiful stories
to Gibbons the reporter did his best to reciprocate, and
incidentally give his story a happy ending by finding
someone who would finance the need. Very soon he discovered that Charles Page was the most dependable source
of immediate relief. But Page, interested in the individuals he helped, shrewdly aware that charity could be over-
done or imposed upon, found time to make some investi-
gations. He learned that financing widows and fatherless
children back to Pennsylvania, Texas or California by no
means covered the need. There were plenty of needy
widows and children who had no place to go. And thus
was conceived the Sand Springs Home.

Page, out of an oil-made fortune which by that time had
grown into several millions, purchased nine thousand
acres of land at Sand Springs, about seven miles from
Tulsa. In the early summer of 1908 he moved a dependent widow and several children into temporary quarters
there. Nine thousand acres may appear to be rather a
large area for a widow to use as a kitchen garden. It was
in fact a measure of the breadth of Page's plan.

It was the philanthropist's purpose to provide an endowment which would be as safe, permanent and free
of fluctuations in income as human ingenuity could de-
vise. The nine thousand acres in and around Sand Springs
settlement was selected with that purpose. It comprised
woodland, pasture land and fertile farm land in variety.
It also provided potential sites for what he purposed to

build into a diversified industrial colony as an added source of income under the endowment and a hedge against fluctuations of the farm income.

With the project that far advanced, Page built the first substantial dormitory for the widows and children whom he intended to make his beneficiaries. It was completed in 1909, and in that same year filled to capacity when the Cross and Anchor Home for Dependent Children in Tulsa was closed. Page welcomed the whole lot of youngsters to his Sand Springs Home.

Then he went to Captain Breeding, the Salvation Army officer whom he had come to know and trust implicitly in the first days of his charities. "Captain," said he, "how would you like to devote all the rest of your life to the welfare of a group of humanity confronted by serious need through no fault of its own?"

"I'd like nothing better," said the Captain. "But to give the task the best that is in me I should need to feel that my own passing years would be provided for until the end."

"We can do that," said Page. "Come to Sand Springs and take full charge of the Home there. Your needs will be provided for under the same endowment which assures provision for the Home, many years farther into the future I hope than either of us can see."

"I'll go; happily and eagerly."

When I met Captain Breeding, nearly thirty years later, there could be no doubt that Charles Page's wisdom in the selection of a superintendent for his philanthropy had been as keen as his business judgment in the arrangement of the endowment funds. Slender, quiet, kindly, with a face refined to asceticism by a lifetime of unselfish practical service in behalf of others, Captain Breeding welcomed his guests with unobtrusive dignity and unaffected cordiality.

But it was when he made his way down the great living

room of the main building of the Home, filled but not crowded with children of all ages, that the extraordinary fitting of the man to the job was revealed. It was a normal Sunday evening gathering of a family and friends. Two or three small groups of youngsters sat upon the thickly carpeted floor playing games as children of similar age doubtless were playing games in thousands of normal American homes throughout the nation. A larger group in a far corner was as happily engaged in an orchestral concert directed by their music teacher. A few little girls in the starched importance of Sunday frocks displayed in honor of "company" occupied davenports between the wide windows.

As the Captain walked the length of the great room a boy of ten or thereabouts in one of the game-playing groups on the floor appealed to him for decision on a dispute which had arisen. He knelt on the floor among the boys, heard both sides of the argument and gave his opinion. It was accepted with a shout, and the game proceeded. He watched for a moment and moved on. Three little girls sprang from their places upon a davenport. One seized his hand. Another put her arm around him, and a third amid much childish laughter pushed him back to a seat among them.

That was Captain Breeding. That was the impression of normal, happy home life which he, under the wise endowment of Charles Page, has been bringing to hundreds of children through nearly a third of a century.

The Home itself has expanded and improved through the years. The original plan of endowment provided for development of industrial enterprises near the Home in addition to the farm. Page built a steel mill, a cotton mill, a glass factory, and other industrial plants, and a railroad to connect them with main lines at Tulsa. A board of five trustees was set up to manage the various features of the enterprise. Its members were selected with as great care and understanding of their business ability and honest interest in the entire project as Captain Breeding had been selected for the management of the Home. Substantial salaries were specified, with a provision that the trustees should devote their business hours exclusively to maintaining and increasing the income from the endowment. Vacancies due to death or resignation were to be filled by appointment by the Grand Master of Masons in Oklahoma.

The Grand Master hasn't had much to do as yet. Most of the original trustees are still active. They do not resign. The jobs are good jobs, and the rewards are great, beyond the independent incomes which they assure. Each trustee has his distinct duties and responsibilities. One manages the railroad which brings in a profit of about $100,000 a year. One directs the diversified farming interests which have been extended from the original nine thousand acres to a total of twenty-five thousand acres of model farms not only in Oklahoma but in neighboring states. One oversees the endowment's interest in the factories which in later years have been leased to experienced independent manufacturers as being more profitable and less burdensome than direct operation. One manages investments in bonds and other securities. All together they meet each Monday morning and listen to Captain Breeding's report on the problems and needs of one hundred children in the Home, and another hundred who live in independent cottages with their widowed mothers, equally under the provision of the endowment. The trustees' responsibility is to see that the needed funds are provided. They take a keen interest in expenditures but do not dictate them. That is Captain Breeding's job.

That it is a job well done along the far-reaching lines originally laid down by Charles Page is effectively demonstrated in a visit to the Home. One of Page's first and most important convictions when he began to plan was that families should be kept together. He discovered that single orphans stood a much better chance of adoption into good homes than two or more children together. So he ruled that preference should be given to larger family groups. Through the many years of development and improvement of the Home this has been one of its outstanding policies.

Orphaned family groups ranging as high as eight in number gain entrance as a group. In the wide and airy dining room such family groups are seated at individual tables. Every effort is made to maintain their family integrity. The oldest child in the group is impressed with the responsibility of helping to guide and teach the younger. Loyalty and solidarity is impressed upon them. Mixed smaller families dine at small tables for six or eight under the eyes of matrons, assistants or heads of departments.

Almost all the food consumed in the home, with the exception of sugar, coffee, spices, etc., is produced upon the Home farms. It is substantial and appetizing food in great variety and unlimited quantities. It is perfectly cooked by an efficient staff and daintily and efficiently served by the older girls of the Home who have not yet found their way into business or domestic life of their own.

All the children received by the Home are committed to it by court action in cases of need. The foundation requires only that they must pass reasonable mental and physical examinations to avoid ill effect upon other children already there. Every effort has been made, and made effective, to avoid any stigma of the conventional old-fashioned orphanage. There are no uniforms and no regimentation. The children attend the regular public schools of Sand Springs. They engage in all the school activities on the same footing with the children of the factory employes or railroad workers or shopkeepers whom they meet there. Those who show special inclination or talent are provided at the Home with additional instruction in music, art or what they will.

The boys are provided with special facilities to find a congenial and productive place in life through trades, farming or preparation for professions. The girls who show facility are taught to make their own clothing, to cook and to do all the normal tasks of housekeeping.

None is ever thrown into the world unprepared to make a living any more than a normal child of a normal family is thrown out. If they want to be office workers they are prepared for office work, and usually they are so well prepared that jobs are easily found for them. If they lose their jobs for any reason they can always go home to Sand Springs. If they show sufficient aptitude, ambition or industry for professional careers they will even be sent to college. All the rules of the Sand Springs Home are designed to make it a home in truth.

It has in fact some advantages which the older children in overcrowded families might well envy. The older girls, for example, have a large and comfortably furnished living room where the little ones are not permitted to intrude. Each has a room of her own, small but comfortable, which she is permitted to decorate in any manner that her fancy may suggest. When I was permitted to glance into a few of the rooms temporarily unoccupied I noted that rotogravure pictures of Clark Gable and Robert Montgomery were much in evidence on walls and dressers.

All the children are provided with regular allowances of spending money, very small for the very young, and somewhat larger for those who may know better how to spend it. They are permitted to go to the moving pictures in the village, and are allowed the other normal entertainments of normal children. Virtually the only punishments for infractions of the few and simple rules of discipline necessary to the management of one hundred children and adolescents under one roof are the restriction of these privileges.

The beauty of the whole system is that it works. Children brought to an age of intelligent thought, independent action and ability to make their own way in the world have been going out from the Sand Springs Home for a quarter of a century. And always it remains home to them. The annual homecoming parties are gay and impressive affairs. At a recent one there appeared among the homecomers a man who had been one of a family of eight which had grown up under the sympathetic and discerning eye of Captain Breeding in the first decade of the institution. Proudly he introduced to the Captain eight children of his own.

"That's fine, Bill," said the Captain. "We're proud of you too."

To me the Captain admitted with a smile that seemed a bit wistful, but with a candor and modesty which proved the strength of his conviction, that in spite of all the best efforts of the directorate the Home could not quite take the place of an independent, self-contained home under the personal supervision of a mother. It is a conviction based upon thirty years of unfailing unselfish effort and close observation of many hundreds of case histories. "In spite of all we can do," said he, "I must admit that the children who are reared in their own family groups under the care of their mothers have made a little better record than those we love and guide with every other advantage here."

That leads to another phase of the far-sighted wisdom of Charles Page's philanthropy. He believed that all children should have the advantage of their mother's care whenever possible. To provide for that he established a widows' colony as an adjunct of the Home.

Widowed mothers who were confronted with the trag-

The late Charles Page, hard-headed and soft-hearted business man, whose rise from a boyhood of poverty to a manhood of wealth and power in the Oklahoma oil fields was crowned with one of the finest self-perpetuating philanthropies ever conceived.

edy of seeing their children taken from them and scattered were provided with homes built to their needs in a section of the foundation grounds. There, assured of all necessities, they were free to rear their children happily together. Such mothers were helped to find jobs. They could maintain their self-respect, and help to provide for their families, relieved of any demoralizing fear of separation or abject poverty. A day nursery cared for the little ones while the mothers were away at their work. The children of school age attended public school exactly as the children of completely independent families in the village attended. Aside from a broad supervision of health and care for the necessities, the Home attempted no dictation over the widows' management of their children.

Page believed with all his heart in independent unregimented work. He had an abiding faith in the opportunities of Oklahoma. An incident of a visit to the Home by William G. McAdoo, now a U. S. senator from California, is illustrative. Mr. McAdoo was being conducted through the service quarters and storerooms when he noted an impressive display of clear golden honey that would have done credit to a national exposition.

"Honey!" he exclaimed. "I didn't know that Oklahoma produced honey like that."

"Yes," said Page, smiling tolerantly, "we produce honey. It's quite simple. We just take a bee and say, 'Here, bee; here's Oklahoma. You're a damn poor bee if you can't make a living and turn in a profit here.' "

That was more than a decade ago. Charles Page died in 1927. The fortune which he made from Oklahoma oil is still turning in a profit of opportunity and happiness to hundreds. The Sand Springs Home Interests form a unique monument to one success in the oil industry.

Fortunately for the sake of variety and the solution of the many problems of modern life the character of men in the oil business has been as varied as the diversity of their products. Take a look at Bill Skelly.

When I met him in his big substantially furnished office in the Skelly Building in Tulsa he was busy chasing a folded pad of paper matches around the top of his desk, attempting to strike a light with one hand while he held a telephone instrument with the other. He was arguing heatedly with someone in Oklahoma City as to whether "Alfalfa Bill" Murray would be the proper man to introduce Alfred Landon to the electorate of Oklahoma. Finally he gave up the pursuit while continuing the argument. He seized the fugitive match folder with his free hand, fixed it between his knees, struck his match, lighted his cigarette and laid it down on a tray where it was permitted to burn itself out.

It was hard to tell whether he was more distressed by the problems of his responsibilities as Republican National Committeeman for Oklahoma, by the fiendish perversity of the slipping match pad or the stab of pain in an injured back every time he moved in his big chair. Whatever the cause of the distress, at the moment he gave the impression of a round-faced eight-year-old boy who has been sharply reprimanded for a misdeed which is a complete mystery to him. And Bill Skelly is fifty-nine years old. No one would suspect it from his appearance, his expression or his superficial actions—until the word "oil" is mentioned.

It doesn't take long then to understand that oil has been the inspiration of most of his tremendous energies for forty-odd years. For Bill Skelly was born to the oil business. His first jobs while still a schoolboy in Erie, Pennsylvania, where he was born in 1878, were as a sort of supercargo for his father in the business of hauling oil-well supplies from Erie to the Pennsylvania fields. He sold newspapers on the streets of Erie while he was still in school and graduated from Clark's Business College at the age of fifteen, but oil was his love and his inspiration.

As soon as he was out of school he "hopped a freight" for Oil City. There he promptly found a job with the Oil Well Supply Company. Enlistment in Company D, 16th Pennsylvania, and service in Porto Rico in the Spanish-American War was only a temporary check to his practical oil education. It was an education complete in every detail of the business, as laborer, tool dresser, driller, scout, leaser, superintendent, promoter, salesman, refiner and business organizer. It included a year as manager of a gas company in Gas City, Indiana, and several years in varied oil activities in Indiana and Ohio. Four years' experience in the opening Illinois field followed, and another four centering around the Electra field in Texas. The Healdton discovery and excitement brought him to Oklahoma in 1914. By that time he had more experience and a broader view of the industry than a hundred average men who might have been making their living from some specialized part of the business through those same twenty years.

He had very little money, but he had a way about him, a way which has been attributed by some observers to the Irish intuition and Scotch caution that were his heritage. It proved to be a way which brought men to his loyal support and money to his investment projects.

He saw and grasped the possibilities of the Oklahoma fields and the steady rise of Tulsa as oil capital of the world, but he was not restricted in that vision. He was equally well acquainted with Texas and Kansas. He maintained and extended his interests there. In the El Dorado field of Kansas he became one of the leading independent operators, organized the Midland Refining Company, and constructed its great refinery. Back in Texas he extended his interests in the Panhandle fields.

By that time, at the end of the World War, Bill Skelly's interests were too wide and too varied for any one man to manage. Incorporation of the Skelly Oil Company was the logical step. In due time it absorbed the Midland Refining Company, and extended and diversified its holdings and activities throughout the mid-continent field. Buying, leasing, drilling, producing, transporting, refining, marketing, the Skelly Oil Company has developed into an enterprise known to virtually every person in a dozen Mississippi valley and western states who knows that oil and gasoline are necessary to the driving of an automobile.

Bill Skelly personally is somewhat of a hero to thousands of college youths and co-eds and graduates for whose edification he endowed the great Skelly Stadium at Tulsa. He is known to virtually every important oil man in the world through his activities in the advancement of the International Petroleum Exposition in annual session in the world's oil capital. He is, or has been, a director of the International Combustion Engineering Company, the Empire Trust Company of New York, the First National Bank of Tulsa, the Oklahoma Iron Works and the Tulsa Chamber of Commerce, and Republican National Committeeman for Oklahoma. It was almost exclusively on the latter job that he was concentrating when I tried to talk with him about oil.

"What's going to happen in the oil business when men like you who have grown up in it and developed it are gone?" I suggested as a lead.

"Plenty of smart and competent boys growing up in it," he said. "Every industry today is a training school where the man below is being trained to become the man above."

The telephone rang. Bill Skelly seized the instrument and found himself in the midst of a new argument on political policy with someone in Oklahoma City. As he talked he fumbled a new cigarette out of a crumpled package, twisted a match out of the pad and began to chase the pad around the desk in vain effort for a light. At last he brushed the folder to his lap, gripped it between his knees, struck the match, took a single puff from the cigarette and laid it aside to burn up while he finished the phone talk.

"You've built up a tremendous business." I tried to get back to oil when he had cradled the instrument. "How do you manage to handle it and do so many other things of public interest on the side?"

"Organization," he said. "It is a big business. Oil is a grand business. Some one ought to write it, and do it right. No one can tell 'em about it. They've got to see it, and feel it. I'll tell you what I'll do. I'll have one of our men take you out through our fields and refineries, and all around. See the whole business. Eat with the men in the field, sleep in the bunkhouses, get acquainted. Damn that telephone!"

This time it appeared to be a call from California, but again with reference to politics, not oil. It was quite evident that Bill Skelly was up to his ears in the national campaign. Again he fumbled out a cigarette and unconsciously repeated the little comedy of the match chase, only to lay aside and forget the burning cigarette.

"You see Way, manager of our radio station in the Philtower Building," he turned to me. "Tell him I'd like to have you get hold of those old-timers' letters that come in each year when we're getting up the programs for the Petroleum Institute. Some of 'em are rich. There's no business in the world like oil. Unless it's politics." He grimaced whimsically or with a new stab of pain in his wrenched back as he reached again for the insistent phone.

Again I watched the W. C. Fieldsian comedy of match and cigarette, and decided to withdraw temporarily in favor of Alfred Landon. This was not Bill Skelly's day to be interviewed on the history of oil in Oklahoma. The spirit was willing but the flesh was weak, and the Republican National Committee was too persistent. He realized it also and came to his feet with an effort and an apology as I arose at the renewed call of the buzzer.

"I'm sorry," he said. "You tell Mr. Way and our public relations department to give you all the information you want. I'll arrange for that trip through the fields." He reached again for the telephone. As I turned at the door I saw him fumble once more for a cigarette and match, but there was no fumbling for words in his answer to some inquiry from Washington.

That was Bill Skelly after some forty years in the oil industry. I did not try to see him after the election. I suspected that the pain in his back might be worse. And from the publicity director of his company I learned that if I made the personally conducted tour of the company's properties it would take me into eleven states where the company owned 1,300 producing wells, held oil leases on 1,000,000 acres, field storage capacity running into the millions of barrels, refineries and transportation facilities to match, hundreds of modern filling stations, and other profitable interests too numerous to mention.

It was more than I felt competent to cover. It had taken Bill Skelly about forty years to understand it. I could not hope to do so in forty days. Oklahoma alone had more interesting stories and characters based upon oil than could ever be crowded into a single book. To my lay mind the men and their accomplishments outweighed in interest the statistics of their accumulated wealth.

CHAPTER XV

ACCOMPLISHMENTS MARK CONTRASTS

IN NO great industry which has played a part in the building of the United States have there been more valid stories of poor boys who became famous than in the oil industry. In no other industry have there been more pleasant tales of honesty and exciting tales of fraud, or of economic and legal battle. A brief glimpse of a few seems necessary to suggest the broad picture.

Harry F. Sinclair, for example, began his working life as a drug clerk and pharmacist. That fact was given nationwide publicity when he was assigned to work in the pharmacy of the District of Columbia jail while serving a sentence for contempt of court in connection with proceedings against Secretary of the Interior Albert Fall on charges of bribery growing out of the Teapot Dome oil scandal a few years ago.

It is not so widely known that Frank Phillips, head of the $375,000,000 Phillips Petroleum Company, began business life as a barber. Frank Phillips sold a hair tonic of his own manufacture many years before he produced or sold a barrel of oil. The start of numerous other captains of the great Oklahoma oil industry was similarly humble. Oklahomans interested in the state's vast oil production, as are most Oklahomans who are not struggling to wrest a living from the surface of the soil, delight in tales of the spectacular rise of the men who have attained the greatest success.

Many of the stories are told with more gusto than documentation. At least a dozen men, active in the oil business, told me that Harry Sinclair's first success was made possible through acquisition of initial capital from accident insurance collected when he shot off a toe while on a hunting trip. I searched long for a quotable source of that statement until I found it as an item in a series of copyrighted stories published in the Oklahoma City *Times* in April, 1934, under the by-line of Harold L. Mueller. So far as I know, Mr. Sinclair has never denied it. Be that as it may, he was on the scene, prepared with intelligence, energy and ambition, if not with capital, to take advantage of all opportunities in the oil business when they appeared.

Harry Sinclair was only twenty-one years old when the first producing oil well was opened at Bartlesville. As a traveling salesman for a wholesale drug company he was familiar with the geography and small towns of Oklahoma and Indian territory. As a resident of Independence, then the center of the Kansas oil industry, he knew something about the business and its possibilities. He was quick to invest what little money he could command in Bartlesville to take advantage of the first oil excitement.

Patrick J. White, representing the Oil Well Supply Company of Pittsburgh, with western offices in Independence, was one of the first outsiders in Bartlesville to take advantage of the business possibilities based upon oil. Sinclair and White were acquaintances. Quickly they became friends. Soon they became associates. When the first Bartlesville excitement ended abruptly with the capping of the Cudahy company's well because of lack of transportation for the meager oil supply and the cloud of uncertainty with reference to the validity of leases of Indian lands, neither Sinclair nor White abandoned hope of eventual profits.

When Bartlesville production was renewed Sinclair and White were there. When the first small boom developed at Tulsa with the opening of the Sue Bland well at Red Fork, Sinclair and White were quickly on the scene. Riches were still far away, but oil was coming closer and closer. When Pat White heard of the first good well at Flat Rock Creek, near Tulsa, he promptly negotiated numerous leases in that area. Sinclair was in Tulsa on business.

White appeared early in the morning at the Robinson Hotel and paced the hallway waiting for Sinclair to appear. Early rising, it appears, was not one of Mr. Sinclair's requisites for success. Also, if this tale can be believed, even his friend Pat White hesitated to intrude upon him in the hours of his beauty sleep. In any event, White was waiting at the door when Sinclair finally appeared to make his matutinal visit to the only bathroom on the floor.

"Harry, I've been walking the floor for hours waiting for you to get up," White fairly shouted.

"What's up? What's on your mind?"

"They've got oil in that well out on Flat Rock Creek," White exclaimed. "I've got a bunch of leases lined up but I've got to have twelve or fifteen thousand dollars to close the deal. Are you in?"

"Let's go out and have a look," said Sinclair. "If it's good I can raise the money."

It was a long trip of three or four miles in a buckboard, but the men made it in record time of half an hour.

"That," Sinclair was quoted many years later, "was the beginning of the White and Sinclair Oil company, which at one time was the largest independent oil-producing firm in the world."

In telling the story to Mr. Mueller of the Oklahoma City *Times* in 1934, Mr. Sinclair added a wistful commentary that "those were the days when you could take a man's word for a thing as being as he said."

That, I have noted in interviews with numerous active and prominent men still associated with the oil business in Oklahoma, was a condition which many of the old-timers still look back upon with regret for its passing.

The late Thomas D. Slick, widely known in the early days as "Mad Tom" Slick because of his wildcatting proclivities, one of the most dynamic, popular and successful operators in the mid-continent fields, was an outstanding example of that epoch of honesty. Claude Barrow, oil editor of the *Daily Oklahoman,* tells me that Slick's estate has been cheerfully busy for years since his death at the task of acknowledging and paying for numerous verbal leases and memoranda scribbled on the backs of envelopes, just as Slick himself invariably made good on all such agreements in life.

Old-timers are proud of the casual manner in which verbal agreements sometimes involving millions of dollars were kept. Murray M. Doan, one of the grand old men of the industry in Tulsa, for example, told me that on one occasion while he was active in the Gulf Oil Company he arranged a sale of one million barrels of oil in storage at $2.50 a barrel, and completed the transaction without the signing of a single paper other than a check for $2,500,000 which was handed to him. There were no delays, no formalities and no difficulties. According to Doan there was, and still is, a certain sentiment in the oil business not apparent in any other line of important industry. A sort of college fraternity loyalty, according to the sentimentalists, pervades the industry.

Josh Cosden told me that he once agreed to pay $12,-000,000 for certain properties, and, in the face of an offer of $15,000,000 from another source for the same holdings before the deal had been put in writing, the original promise was made good to him at the $12,000,000 figure.

Harry M. Crowe of Tulsa adds his testimony to the same effect, citing one example in which Tom Slick paid $30,000 for one block and $50,000 for another block of leases without even a penciled memorandum on the agreement. In each case legal title to the property involved was handed over in due time.

Crowe tells another story which illustrates not only the strict adherence of oil men to their pledged word on any business deal, but the speed with which news of oil discoveries spread throughout the country and the promptness with which men in the business sought to cash in on the news. When a well on a small lease in the early days of the Little River Pool came in with alluring promise, Crowe managed to buy the royalty rights on five acres in the neighborhood. He had hardly reached his office after a price of $20,000 for his purchase was agreed upon when a phone call from Los Angeles informed him that a man on the Pacific Coast was offering $30,000 for the same royalty rights. But there was no attempt on the part of the original owner of the rights to welch on his $20,000 agreement with Crowe.

Despite sentimental memories to the contrary, not all the oil men were members of that lodge. Harry Campbell, a pioneer of Tulsa and a one-time judge, tells of one deal in which not one but two parties welched, with the loss of nine-tenths of a stipulated $10,000 fee to him. It happened in one of the early periods of heavy production when the supply so far exceeded the demand that oil prices were below the cost of production.

According to Campbell's story, Charles N. Haskell, first governor of the state of Oklahoma, contracted to buy one million barrels of oil held in storage by a prominent independent producer. The price was to be only twenty-six cents a barrel. But before delivery started the Constantine Refinery, now defunct, offered thirty cents a barrel for the same oil. The producer refused delivery to Haskell at the lower price, and made delivery to the refinery at the higher price. The difference, representing a clear loss to Haskell, was $40,000. Haskell engaged Campbell as attorney on a twenty-five per cent contingent fee to sue the welching producer for the $40,000.

Campbell says he was in a fair way to obtain judgment for the entire $40,000 when he was privately informed that Haskell had settled with the producer out of court, and without the advice or approval of his counsel. Campbell was never able to learn precisely what Haskell obtained. He remembers distinctly, however, that he himself received only $1,000 and a hard-luck story from his client instead of the $10,000 which should have been his if the suit had been carried to its logical conclusion.

Other men, especially younger men, in the industry are inclined to jeer at many of the stories of straightforward honesty and devotion to the sanctity of verbal contracts in the history of the oil business. Authorized signatures on properly drawn agreements duly placed in escrow are essential, they insist.

Probably both attitudes have been justified in their time. The older men who have mixed sentiment with their memories of the great days of active development doubtless have valid reasons for their treasured belief in the sanctity of a man's word—win or lose. But when the profits began to run into the hundreds of millions of dollars, when impersonal corporations rather than well-known individuals became involved, when lawyers began to point out the possibilities of profit in lawsuits, and demonstrate their points, the situation began to change.

That has been the history of similar development on other frontiers, especially frontiers built upon exploitation of mineral rights. In the first rush to the California gold fields half a century before the first commercial well was opened in Indian territory, crime was practically unknown. All a miner needed to do was to leave his pick and shovel on his claim to establish a title which was not questioned. His personal property, sheltered only by a tent or a brush hut, was inviolate. But that happy condition faded very quickly after the riffraff of the world began to arrive on the scene, and laws and lawyers came into action. The same was true of the opening days of the great mines of Colorado, Nevada and Montana.

As recently as 1901 the late Jim Butler, discoverer of the great silver and gold deposits of Tonopah, Nevada, demonstrated the inherent honesty of a pioneer people by granting scores of leases on the main vein of Tonopah with nothing more than an oral agreement in each case. Every lease was worked to the date of expiration and every dollar of royalty paid without one lawsuit being filed in that period. Such common honesty and loyalty to a pledged word seems to have been a pioneer principle. Doubtless it existed, as numerous old-timers testify, in the opening of Oklahoma's first oil fields. But not for long. It passed with the passing of the frontier. Those men who would have continued to follow the practice were forced by the devious practices of a more civilized era to adopt other methods in self defense against the greed of mankind.

The devious practices were by no means confined to the corporations or individual white men. Some of the Indians were as wily as any white. The so-called Tommy Atkins lease is a classic example of Indian guile and chicanery which eventually involved millions of dollars. Half the old-timers in the oil business are familiar with the general outline of the story, but I questioned a dozen before I met one who could give me details from his personal experience. That one was Murray M. Doan.

At the time of the incident Mr. Doan was active in the affairs of the Gypsy Oil Company, then an operating subsidiary of Gulf Oil. According to Mr. Doan, Harry Bartlett of Sapulpa brought in to the Gypsy a lease on 160 acres in what later became the Cushing oil field, one of the richest and most famous in the entire state.

Preliminary investigation indicated that the acreage specified in the lease was an allotment to one Tommy Atkins, an Indian boy. The lease was signed by Minnie Atkins, Tommy's unmarried mother, who was said to have inherited the property from her son. It was properly drawn and witnessed. The woman's authority appeared to be sufficient, and the title valid.

Gypsy opened a highly productive well on the lease. But the boom in the Cushing field was under way, and two other companies questioned the validity of Gypsy's rights under the lease. Further investigation, precipitated by this cloud on the title, revealed that the lease was a forgery. But the Gypsy had proved the high value of the property. It had invested its money and it had some claim to consideration. It sent out searchers for Tommy Atkins. The vague trail led all over the United States and parts of Mexico and Canada. It was a tedious and expensive job, but oil property worth millions was involved. And at last Minnie Atkins was found cooking in a camp in the Northwest.

Without much argument she was induced to sign a written statement which only partly clarified the situation. According to that statement there never had been a Tommy Atkins, or at least she had never had a son named Tommy Atkins. But when the formality of an oath to the accuracy of the statement was invoked Minnie became so impressed with the importance of the document that she refused to give it up.

In the meantime more and more oil and money were flowing from the Cushing field. It was a prize growing richer every day. The litigants who sought to take control of the Tommy Atkins lease from the Gypsy Company brought witnesses to testify that there was in fact a Tommy Atkins, but that he was the son, not of Minnie, but of her sister Nancy. They added that Minnie had borrowed the boy from Nancy to pose as her son in the granting of the original allotment, before oil was suspected there.

Gypsy realized that its legal claim to the property was extremely tenuous. Charles Page, whose ultimate accomplishments have already been recorded, induced the Gypsy to abandon its claims in consideration of reimbursement for the money actually expended in development and litigation. Shortly afterward Page induced the other litigant to withdraw. The Tommy Atkins lease went into history as a complicated fraud perpetrated by the Indians. But with title finally established, the oil from that lease added to a fortune of millions which Page eventually used to far better purpose than many oil-made fortunes have been used.

Before Page had accumulated his score of millions the oil riches of eastern Oklahoma were having the usual effect of riches upon politics and economics as well as upon the social life of the people. One of the first battles to flare from that development came between Governor Haskell and Charles West, first attorney general of the newly admitted state.

That battle arose out of a situation which developed with peak production of 117,440 barrels of oil daily in the Glenn Pool in 1907. That tremendous production of oil from a small area brought about a serious problem of transportation and a break in prices to as low as thirty cents a barrel. Some independent producers were unable to sell their oil at any price. They could not finance the construction of proper storage tanks, and vast quantities of oil were run into hastily constructed earthen reservoirs where quality and value deteriorated rapidly. Oil from other wells was simply allowed to run to waste down the creeks. Fires were an incident of daily occurrence throughout the field. Rolling billows of oily black smoke obscured the sun. The waste was beyond comprehension. It was a scandal, and to many of the producers and owners of property under lease it appeared to be a crime.

President Theodore Roosevelt's popular "trust-busting" campaign of the same period had made all newspaper-reading Americans trust-conscious. Oklahomans were no exception. Independent producers were quick to blame conditions upon what was then generally recognized as the most effectively organized, powerful and ruthless trust in the world—John D. Rockefeller's Standard Oil Company. They saw vast fortunes thrown to waste, great riches snatched from their hands, unlimited possibilities of wealth and development of a great new state arbitrarily restricted by the greed and power of Standard Oil.

The hand of the Standard Oil Company appeared in the eastern Oklahoma field, especially in the spectacular early production of the Glenn Pool in the promotional work of the Prairie Oil & Gas Company. Most important of the P. O. & G. activities in so far as the independent producers were concerned was the pipe-line construction which promised to provide an outlet for oil to market. But before the projected line could be completed, independents who had been denied transportation locked horns with the trust. Simultaneously Charles West, Attorney General, and Charles N. Haskell, Governor, became involved in the controversy on opposite sides and raised it to an issue of major importance and publicity.

Independent producers complained bitterly that the Standard Oil Company, as represented by Prairie Oil & Gas, discriminated against them, refused to transport their oil, or even to buy it at the absurdly low price of thirty cents a barrel. Attorney General West set out to block further pipe-line construction of the P. O. & G. by injunction proceedings technically based upon the ground that it was a foreign—i. e., not an Oklahoma—corporation, illegally disrupting communications within the state by digging its trenches and laying pipes across state highways.

At the same time Governor Haskell had abruptly reversed an attitude of opposition to the Standard Oil Company in which he had declared that organization to be the source of most of the ills of mankind as then revealed in Oklahoma. Haskell appeared to be all in favor of the Prairie Oil & Gas pipe-line development. His main point of contention, however, was a technical one of jurisdiction.

The Governor, not the Attorney General, he maintained, had the sole authority to initiate such action as West had initiated against the "trust."

The battle raged, while oil development was virtually stopped for the time. Omer K. Benedict, editor of the Oklahoma City *Times,* and a man whose name is still honored by many of the leading old-timers throughout the state, put popular suspicion into print in an editorial published early in August, 1908.

"He (the Governor) says he is holding office and the people have the right to know what he is doing," Benedict wrote. "That's true, Governor. What were you doing at the Coates House, Kansas City, Mo., on the 16th of June, in company with Mr. O'Neil of the Standard Oil, and what were you doing in Independence, Kan., on the 17th with Mr. O'Neil, when you were supposed to be in Muskogee? And please tell us whom you were supposed to meet in Chicago."

Governor Haskell promptly answered that intimation of a sell-out to Standard Oil by bringing about the arrest of Mr. Benedict on a charge of criminal libel. All Oklahoma was on edge. The extent of the state's excitement over the Haskell-West battle on the issue of Prairie Oil & Gas pipe line development was revealed in two red-ink headlines running across the entire front page of the Oklahoma *State Capital:*

ATTORNEY GENERAL WINS—HAS RIGHT
TO BRING SUIT FOR STATE—GOVERNOR
    USES LAST CHANCE TO USE POWER

The story was based upon a decision by Judge A. A. Huston upholding the Attorney General's right to bring legal action against the oil company despite the Governor's contention to the contrary. Haskell promptly countered by obtaining from the State Supreme Court an injunction to block further prosecution of the Attorney General's suit.

"Haskell determined to be King of Oklahoma," read one headline covering a phase of the story. "Oil Men Are Disgusted," read another. "Claim They Have Had Enough of Double Cross and Want the Officials to Stop Quarreling While They Suffer."

And in the course of time they did stop quarreling, though not until a new state law specifying more clearly the rights and responsibilities of oil and gas companies had been adopted. Following elaboration and clarification of that law, drawn by the Attorney General himself in 1910, John D. Archbold of the Standard Oil Company, accompanied by counsel for the company, came to Attorney General West and agreed to discontinue discrimination against independent producers. Mr. West told me about it twenty-six years later in his home in Oklahoma City.

The Prairie organized a local company, bringing it clearly within the jurisdiction of Oklahoma law. Then it completed its pipe-line service, and the more spectacular features of the struggle were over. Other pipe lines extended steadily. The way was cleared for a new era in the development of Oklahoma oil fields.

CHAPTER XVI

BANKING THE OIL GAME

No such epochal events as those of the modernization of the last frontier into a region priding itself upon cities suggestive of New York in miniature could take place without simultaneous development of banking facilities. With such men as T. N. Barnsdall, H. V. Foster, Harry Sinclair and Frank Phillips actively engaged in the development of Oklahoma's first great oil fields, advancement was inevitable. With such companies as Standard Oil, Indian Territory Illuminating Oil, Prairie Oil & Gas and Gulf Oil in action it was inevitable that all the machinery of great corporations, good or bad, should become an effective part of the general progress.

No one will maintain that the power and practices of the great corporations, often ruthlessly applied, were always beneficial, although they did produce the oil and help build the cities. At times they were vicious. At times they were illegal. An example of the latter is in the record in the case of the State of Oklahoma vs. Waters-Pierce Oil Company as early as 1910.

Waters-Pierce was sued by Attorney General West on charges of violating state laws in various oil development and marketing activities. After some months of litigation, during which evidence was piled high against the corporation, representatives of the company confessed the misdeeds, paid a fine of $75,000, and submitted to an injunction against any further action in restraint of trade.

The evils of ruthless corporation methods had been widely publicized, but independents recognized the value of effective organization in getting out the oil. One of the chief advantages enjoyed by the larger companies with financial backing in Wall Street, Pittsburgh, and Cleveland was that of flexible banking facilities.

The first banks of Muskogee, Tulsa and Bartlesville were ordinary small country banks, as the towns in which they were located were ordinary small country towns. The bankers knew enough about their communities to serve the local merchants, and the cattlemen who were their most important customers. But a banker who could appraise a herd of cattle and grant or refuse a loan sufficient to carry it through the winter was not so efficient when it came to appraising an oil proposition.

That was the situation when Frank Phillips entered the banking business in the awakening oil town of Bartlesville. Frank Phillips was a product of a frontier which ended earlier than that of Oklahoma. He was born in 1872, the first white child in Greeley County, Nebraska. His father was a pioneer among the farmers in that region from which the Indians had but recently been removed. In those days the farmers who were settling the prairies built their own shelters and fences, and plowed virgin sod for their first crops without help from any government agency. They had a tough time. But they produced men. Frank Phillips was one of them. He was a babe in arms when the first grasshopper scourge swept the new homestead. He remembers hearing his father say that the grasshoppers even ate the fence posts, and that the chinch-bugs were rolling up the wire when the parents decided that they had better leave to save the baby.

The family moved back to Iowa and settled on a farm near the village of Creston. The boy grew up in that hardworking, thrifty, competent environment. He was smart enough to realize quite early in life that the plow and the pitchfork did not open opportunities which appealed to him. But opportunities for smart boys, especially sons of

poor parents, were all limited. The best he could do was
to go to Omaha and learn the barber's trade. That done,
he returned to Creston and found a job in one of the village's two barber shops.

Evidently the Iowa villagers of forty-odd years ago had
some of the same human characteristics, including vanity,
which may be found on Hollywood Boulevard and Park
Avenue still. Some of the young man's customers feared
that they were losing their hair. He mixed a tonic, applied it vigorously and achieved results which encouraged
some of them. The good word spread. The young man
began to sell the tonic, with excellent advice as to massage
of the scalp. Soon he owned the shop. Soon afterward he
owned the only other shop in town. He came to the favorable attention of Creston's banker. The aristocracy of
accomplishment was then the only aristocracy of any importance in that place. The young man learned the funda-
mentals of banking and married the banker's daughter.

News of the new oil riches and possibilities of Oklahoma came to Iowa. Frank Phillips entered the banking
business in Bartlesville. Very quickly the banking business, as personified in Frank Phillips at least, realized its
affinity with the oil business.

Frank Phillips brought in his competent brother, L. E.,
then a coal salesman, to run the bank while he himself
devoted his talents almost exclusively to oil. Another
brother, Fred, joined them. A younger brother, Waite,
whose much-loved twin and inseparable companion had
recently died, was induced to come to Oklahoma and go
to work to build himself a new interest in life.

Out of that banking venture and the oil business into
which it speedily led grew the $375,000,000 Phillips
Petroleum Company of today. Hundreds of thousands of
motorists in the Middle West and South are familiar with
the shield of its insignia announcing "Phillips 66" gaso-

line and lubricants on innumerable filling stations in a score of states.

Frank Phillips still works as enthusiastically and persistently at his job as he has worked for forty years, as most similarly successful men work throughout their lives. Perhaps half the year he spends in direct management of the company's fiscal affairs in New York, and the other half in equally close attention to the vast business at the home office in Bartlesville.

His chief relaxation is at the ranch which he operates a few miles from Bartlesville. It is one of the finest in Oklahoma. It is not only a ranch where blooded stock is raised and marketed—at a profit—but a vast game refuge and preserve. Countless acres of rolling woodland, brushland and meadow are inclosed by fences which the grasshoppers could not eat nor the chinch-bugs roll. They are high close-woven wire fences on metal posts set in concrete. Behind them, as free to wander and to propagate as upon their native veldt, but with the advantage of convenient shelters and ample fodder in the winter storms, are representatives of most of the horned and cloven-footed beasts of the entire world.

Waite Phillips also has a ranch, or ranches. When a mutual acquaintance introduced me to Waite Phillips in his modern sky-scraping, express-elevatored Philtower Building in Tulsa, I was requested not to ask personal questions, lest the acquaintance be quietly taken to task at a later date. Waite Phillips is a quiet, modest, unostentatious, philanthropic citizen of Tulsa, more universally respected and admired in his home town than any man I have ever encountered in any community in the United States.

In consideration of my sponsor I did not ask him about the ranch which is famous among his friends, and unknown to most of the world. Friends however offered one or two sidelights which give an impression of it. When a group of friends were hunting there as guests of the owner, the Governor of New Mexico joined the party by invitation. In the persiflage of an evening about the great fireplace after a day's shooting the Governor announced that he was considering the advisability of having a bill introduced in the legislature to buy New Mexico back from Waite Phillips. From another source I was told that the ranch is so vast that there is actually a difference of opinion to the extent of eighty-seven thousand acres between the manager and the owner as to the extent of land included. That probably is an overstatement, but it does suggest a large ranch.

Be that as it may, the New Mexico ranch represents only a small fraction of the fortune which Waite Phillips has acquired from oil and real estate in Oklahoma, and which he has poured back into the improvement of Tulsa, which helped to make it all possible, just as the oil made Tulsa possible.

He has built two of the finest, tallest modern buildings of the many fine modern buildings which mark the city. He has given numerous other buildings to eleemosynary institutions. He still owns and continues to improve other city property. He has more or less retired from activity in the oil business, although he still retains an interest in the Phillips Petroleum Company. There is a story about that also.

When his elder brothers, having extended their original interests from banking to oil, sold out most of their oil properties and again concentrated on banking, Waite Phillips stuck to oil. It was as an independent then that he developed his holdings to a point at which Blair & Company of New York could earn a profit by paying him twenty-five million dollars cash and turning the properties over to the Barnsdall Corporation.

The Phillips brothers, Frank, L. E., Fred and Waite, at one time appeared to be practically out of the oil business.

But not for long. When the older brothers had sold their most productive holdings and concentrated on banking they retained some leases on unproved Osage lands. As time went on, extension of development revealed that there was oil in vast quantities under those lands. Willy-nilly, the older brothers found themselves back in the oil industry. Those leases developed into the largest richest oil properties they had ever owned.

Waite Phillips, whose $25,000,000 cash sale was at a later date, had also retained some unpromising properties which neither Blair nor Barnsdall had considered worth adding a million or two to the original check to control. But as the oil fields of the state expanded, with new and rich discoveries each year, Waite Phillips in co-operation with Otis McClintock proved that he had made no error in acquiring the despised holdings or in retaining them. The first thing he knew, he also was back in the oil business as a leading figure in the Independent Oil and Gas Company.

The Phillips Petroleum Company was incorporated in 1917 to capitalize and manage the older brothers' vast new interests in the Osage lands. Later they bought their first refinery and began retailing gasoline. Forward strides from that moment were almost too swift to follow. And when Waite Phillips decided for the second time that he would find more satisfaction in his real-estate and philanthropic activities than as an oil operator it was logical that his Independent Oil and Gas Company should be merged in the Phillips Petroleum Company. The arrangement has given him a little time for himself and a great deal of money for all concerned. It's a grand business—if you know how.

And one of the necessary factors in making it grand is the banking business, as suggested at the start of this chapter. The Phillips brothers were not alone in recognizing that fact very early in the game. Harry Sinclair, starting on even a more tenuous financial shoestring than did Frank Phillips at Bartlesville, quickly realized the necessity of adequate, flexible, sympathetic banking service.

Sinclair and White, along with various other ambitious and more or less independent oil men of the time, found themselves somewhat handicapped when the Farmers National Bank of Tulsa closed its doors in 1909 after the failure of the Columbia Bank in Oklahoma City. The Farmers Bank was, as the name makes clear, primarily a farmers' bank. It had accommodated only such oil men as could convince its agriculturally inclined management that they, personally, or their oil prospects, were sound securities. Some of its total of $400,000 deposits was oil money.

Four hundred thousand dollars was a great deal of money in Tulsa thirty years ago. Its momentary withdrawal from circulation threatened catastrophe. Independent operators were having a hard enough time to finance new leases, drill new wells, and market their product in competition with concerns which enjoyed the comparatively unlimited financial support of the Rockefeller or Mellon interests. The closing of the Farmers Bank precipitated a crisis in the struggle between rich, experienced and ruthless corporations and rising, ambitious, hard-headed individuals.

The independents simply had to have money to protect and advance their affairs. Patrick White, Harry Sinclair, Robert McFarlin, J. H. Evans, F. B. Ufer and half a dozen others managed to scrape together some $400,000. They organized the Exchange National Bank, took over the assets and liabilities of the Farmers Bank and announced that all depositors in the defunct bank would be paid in full—with oil money. The Exchange National was to be primarily an oil bank. Every official and every man on its directorate was personally interested in or familiar with almost every oil well, lease or prospect in the state. When an oil man came to the Exchange National for banking accommodation he was welcomed and served. Usually the values behind any securities which he might offer were known to one or more directors. If they were not known, the facilities for examining them quickly, accurately and sympathetically were available.

If the securities or prospects were sound the needed money was advanced, promptly and generously. Tulsa merchants, local depositors and farmers who had seen their deposits tied up with a probability of almost total extinction in the collapse of the Farmers Bank were kindly disposed toward the Exchange National, which was paying them off without loss. Tulsa itself was convinced anew that a great future as a city lay in bringing more and more oil men to realize that there was the natural center of the world's oil business. Possibly the Exchange National Bank's service to oil men did more than any other one thing to achieve that goal. The bank received deposits from oil workers and oil promoters in every country in the world. Improved and constantly improving hotel accommodations, transportation and office space, and an enthusiastically oil-conscious chamber of commerce, did their part to build and maintain the city on the profits of oil and service to the industry.

With the rise of the Exchange National the independent operators in Oklahoma even acquired some advantage over the far richer and more powerful corporation men, who were forced to report to Pittsburgh or New York. The oil business of Oklahoma, especially in its earlier days of prospecting and wildcatting, frequently required quick and decisive action. A minor but significant example of that has been cited in White and Sinclair's prompt deposit of $12,000 to hold leases taken by White on Flat Rock Creek within a few hours after White saw the first oil flow there. If they had been forced to go to New York to get the money, to report to some impersonal corporation in

Pittsburgh and await a meeting of a board of directors, there would have been a different story.

More than one similar opportunity was made a reality through the prompt and practical assistance of the new oil bank. But it was not all as simple as it sounds. The Glenn Pool's tremendous production in 1907 had slowed down prospecting and development as it cut prices to a minimum. But no oil man or potential oil man could forget the independent fortunes taken from that same pool.

Among the most spectacular financial successes started there was that of Robert M. McFarlin and James A. Chapman. McFarlin, a Texas rancher, had moved his herds into Oklahoma some years earlier because of a drouth so intense that, according to one cowhand, it took all the moisture from a forty-acre lot to rust one nail. McFarlin and Chapman learned the oil business profitably in the Glenn Pool. By the time the Exchange National Bank was organized McFarlin was sufficiently rich and oil-wise to become one of its original directors and stockholders. The spectacular extent of his financial rise will be revealed at a later point in this narrative.

While oil prices were near their bottom and development generally slackening in the year after organization of the Exchange National, a weary traveler applied for shelter one dark night at a small log house near the Cimarron River in Payne county. Within the cabin were its owner, a man named Frank Wheeler, his wife and nine children, of whom eight were girls. They were neither well equipped nor well disposed toward unbidden transient guests.

Wheeler had bought his quarter-section four years earlier, and had found that a man and wife and nine children could not make a living from its sixty acres of plow land and one hundred acres of rocks and scrub. He spent most of his time traveling about the countryside working at his trade of mason when he could find a job. It was a pitiful, poverty-stricken little home.

Visitors were rare in that sparsely settled region. Most of the settlers had more than they could do to feed their own families. The spirit of hospitality is somewhat cramped when the cupboard is bare and most of the household space occupied by children sleeping on makeshift beds. Occasional visitors in the past had identified themselves as oil prospectors, but there was not a producing well within twenty-six miles of the Wheeler farm, and the farmers were fed up with promises.

But this night visitor was a persuasive talker. He was miles from the crossroad settlement known as Cushing. The night was dark. The road was hardly more than a trail through the woods. Frank Wheeler took him in. And when he saddled his horse and rode away next morning he carried with him an oil lease on the 160 barren Wheeler acres. He was Thomas D. Slick, then known as one of the wildest wildcatters in all the Oklahoma country.

Tom Slick had grown to manhood in the oil regions of Pennsylvania. The smell of oil sands was perfume to his nostrils. It told him more of what lay beneath the earth, a hundred, a thousand, two thousand feet, than any aromatic scent of spices could suggest of tropical plantations beyond the seas. He had learned his geology in a practical school. He had learned the hard lessons of human relationships at the same time. He had to know both to get along as a wildcatter in the oil game. Half the time he might be looking for a place to drill, but the other half he was looking for a partner who would finance the work.

At the time of his appearance at the Wheeler farm he was working under an agreement with M. and B. B. Jones, bankers in the little town of Bristow, Oklahoma. The Joneses furnished just enough money to permit him to travel around signing up the farmers on leases. It was a year after the Wheeler lease had been signed before enough neighboring land had been leased to warrant the Bristow bankers in putting down a well. Slick selected a site three miles east of the Wheeler farm. When the well was still dry at two thousand feet the bankers shut off their funds.

There was no oil. But the sand brought up by the drill gave forth an odor to which Tom Slick's nostrils dilated as the nostrils of a cavalry charger to the acrid odor of powder smoke. Eyes and touch confirmed the judgment of his nose. There was oil in them thar hills, but nobody except Tom Slick believed it. Studying the geological picture as suggested in the various rocks and sands brought up through two thousand feet by the drill, and elaborating his mental diagram from his knowledge of the dips and angles of the surface, Slick selected another site, three miles from the first well, on a small creek which ran through the Wheeler farm. But the banker refused to put up money for another wildcat. He still held an interest in many acres of oil leases obtained by Slick, but he preferred to let someone else put up the "dry-hole money" to test their value.

Slick was desperate. But he was also as confident that there was oil under the Wheeler farm as if he could actually look down into the vast body of oil-bearing sands which his calculations had told him was there. The fact that he had been equally confident in numerous other prospects, only to be disappointed and to disappoint his financial backers made no difference. A thorough-going wildcatter must be that way.

Tom Slick called a meeting of the businessmen of the village of Cushing, the nearest trading center and the logical metropolis of any oil pool which might be developed in that area. He told them what he had discovered concerning the geology of that area. He pictured the riches which would come to their crossroads town with the bringing in of a great well on the Wheeler farm. He made a stirring, appealing speech. He asked them to raise only $8,000 for the proposed well, in return for which he would give them a one-half interest in all his lease rights, which would be worth millions when the well came in.

But half the residents of eastern Oklahoma had heard that sort of talk during the preceding ten years. Some of them, trying to turn the dream into reality, had lost their shirts. The Cushing businessmen declined the proposition. Tom Slick departed, and racked his brains for some other financial solution of his problem.

At last he remembered an acquaintance named C. B. Shaffer whom he had known in the oil fields of Pennsylvania, and who had acquired a fortune in the business. Shaffer lived in Chicago. Tom Slick borrowed $100 and journeyed to Chicago. He laid his geological data, plans and hopes before Shaffer. Shaffer decided to risk $8,000 for an interest in the proposed well plus one thousand acres of leases around the Wheeler farm. Drilling started promptly and proceeded swiftly.

On March 12, 1912, the well came in with a roar of gas-driven oil that literally flooded the surrounding earth. The great Cushing Pool had been discovered. Every effort was made to suppress the news until more and more leases had been signed up through the countryside. The well was capped immediately, and men worked feverishly to shovel earth over the oil-flooded area around the well to conceal as far as possible the potential riches of the field.

There were no paved or surfaced highways in the region at that time. None of the scattered poverty-stricken farmers had telephones or automobiles. Normal communication between the country people was slow and difficult. But there is something about oil as there has always been something about gold which seems to give news of its discovery a psychic power of dissemination among the men who seek it. Hardly had C. B. Shaffer arrived at Cushing with J. K. Gano and a corps of experienced lease-getters in

A group of pictures revealing how closely the real estate operators followed the oil discoverers, and their optimistic offering of "business lots."

(Courtesy *The Daily Oklahoman*)

answer to a wire from Tom Slick when other oil men, including independents of every grade and some of the representatives of leading corporations, began to flock into the wildly excited village of Cushing.

The first arrivals rushed straight from the train to the livery stable to hire a rig which could take them to the scene of the strike. And there they found they had been circumvented by the smart Mr. Gano. He had engaged and paid in advance for every horse and rig in Cushing. His lease-buyers were already scouring the countryside. He had even hired all available horses and buggies and buckboards from near-by farmers and placed the animals in a convenient pasture under armed guard.

Most of the men who sought to profit by Tom Slick's discovery found that they must first walk the cross-country miles to the well on the Wheeler farm if they wanted to assure themselves of its reality, and must continue to walk through the surrounding farms and countryside if they wanted to take a chance of finding anyone who had not yet signed away a lease upon his fields and hills.

There was one automobile in Cushing. Its owner had not been included in the Shaffer-Gano roundup of transportation facilities. He charged and collected in advance $25 each for hauling oil men to the well. Leases were signed over a wide area. Tom Slick and C. B. Shaffer owned what appeared to be the best of them. B. B. Jones, Slick's original backer in the lease-buying business in that area, and C. J. Wrightsman, then a rising attorney in Tulsa and now a prominent oil man, had a large interest with Slick in an extensive block of leases. All were made wealthy almost over night.

Big men and little men, rich men and poor men, cleaned up fortunes from the Cushing Pool in the first two or three years of its development. Frank Wheeler, the poverty-stricken farmer with nine children and not enough credit to buy a sack of flour, who had given Tom Slick a night's shelter and an oil lease on his barren 160 acres, was soon receiving an income of $1,250 a day. In three years there were sixteen wells on that farm producing 2,500 barrels of oil each day. Wheeler's income on his one-eighth royalty had stabilized at $300 a day.

No "Coal-Oil Johnny" was Wheeler. He had known poverty and hard work and a desperate need of thrift. He did not intend to find himself again in that distressing situation. He deposited most of the income in banks and expended it cautiously. His first important outlay was for a $15,000 home in Stillwater where an excellent school assured the education of five daughters who were still unmarried. Carefully selected farms were purchased for his older, married daughters, and operated by their husbands. He purchased two automobiles and traveled with pleasure and profit but without ostentation. He enjoyed the mountains and coolness of Colorado with its trout streams and hunting in the summer, and the comforts and entertainment of Florida in the more bitter weeks of an Oklahoma winter.

Aaron Drumright, who owned 120 acres south of Wheeler's, which he had purchased for $1,500 obtained from a lucky deal in the Gregory, South Dakota, land drawing six months before the oil discovery, took out $20,000 cash from the property and bought a farm near Parsons, Kansas.

R. A. Fulkerson, who owned a little land south of Drumright's, and was so poor that the only cash he ever saw was two dollars at a time acquired from the sale of a cord of wood which he chopped and hauled all the miles into Cushing, was able to buy one of the best farms in his native Kentucky and establish himself in comfort and prosperity.

Sarah Rector, a ten-year-old negro orphaned daughter of a Creek slave, who had inherited 160 barren acres originally allotted to her father under the direction of the

Dawes Commission, saw a well come in on her land flowing 3,000 barrels a day. Other wells followed closely.
The child's royalties, under control of a guardian, gave
her a $10,000 home in place of the board shack in which
she had lived in direst poverty. It supplied her with all
the educational advantages which she was able to assimi-
late, and the best clothing, food and servants which could
be found by her guardian.

And so it went. The village of Cushing itself was the
scene of some of the wildest excitement which Oklahoma
had known since the land rushes of the '80's and '90's.
On the day before the gushing of oil from the Wheeler sand
the town had had a population of 300. On the day after
the news had been broadcast throughout the mid-continent oil fields the population leaped to 800. And so it
grew until 6,000 persons overflowed the townsite and the
adjacent pasture lands.

There was no sewage system. There was no water supply except from privately owned wells. There were no
habitations beyond the immediate needs of the town's
people. Wiser and more experienced men among the excited throng had brought tents or blankets or both, with a
limited food supply. More had neglected to do so. A rag-
town of tents, blanket shelters and huts sprang up within
and around the village. As the days hurried by the crowd
increased, and the excitement increased. Hundreds of
penniless prospectors, hopeful of obtaining something,
anything, of the profits promised, jammed the streets,
fought for food at the garbage cans, slept upon the ground.

Gamblers, women, liquor peddlers, vultures, harpies
and the riffraff of the country crowded in. There was no
sanitation, no public health officer, and no adequate police power to enforce health regulations if they had existed. Meningitis, scarlet fever and typhoid began to take
their toll. Crime flourished.

Carl Blackman of Tulsa, who was there, hardly more

than a boy, eager, excited, observant, has given me that sketch of the Cushing boom. Doubtless there. are hundreds of persons still living in Oklahoma and elsewhere throughout the country who could add details and color. It was only twenty-five years ago, and the men who were then battling toward the development and the rewards of the frontier's oil wealth were mostly young men. The life required the stamina, energy and optimism of youth.

Blackman's first visit to one of the few barber shops in the overrun village was greeted with the information that the man who had just left the barber's chair had been shot dead within a few feet of the door. But everyone was too busy or too excited to pay much attention to such incidents. It was a land of freedom. Its people favored no restrictions. A whiskey peddler who had been twice arrested had warned the officer that he would not submit a third time. When the officer attempted the third arrest the peddler shot him down. There was no prosecution. Everyone was too busy.

A gambling house which was one of the first by-products of the money which had suddenly been made available in Cushing was held up and robbed on two occasions. The proprietor then established lookouts, sharpshooters, properly armed, upon a raised platform behind a loopholed screen. A cold-eyed bandit faced the proprietor on the floor with a drawn six-shooter, forced him to call the lookouts from their vantage point without their guns, and robbed the place for the third time. In that spirit Cushing grew and Cushing's fame swept the land.

The publicity which accompanied and assisted in that growth was strange and varied, sometimes favorable and sometimes unfavorable, but always interesting. Accidents in the oil fields, generally due to fires or mechanical misfortunes, had become so common in the many years prior to the Cushing development that they gained little publicity outside the groups immediately involved, but

Cushing managed to introduce something new in the way of accidents.

W. J. Flanagan, Mr. and Mrs. Ora Lyle and Mrs. Lyle's sister, Miss Lulu Reed, were sightseeing through the Cushing field on a muggy cloudy day when suddenly their automobile was wrecked by an explosion of such violence as might have been due to a huge aerial bomb. Mr. and Mrs. Lyle were killed. Flanagan was very severely burned and Miss Reed less severely burned. The precise cause and nature of the accident was a mystery which engaged the attention of scores of engineers and amateur scientists familiar with the field.

It had not been solved when J. W. Witherup and O. T. Flanagan, drilling contractors, driving from Drumright to the well at which they were working, ran into a similar explosion on the highway which took the lives of both men. It was then recalled that atmospheric conditions were the same as had existed on the night of the other accident, with great humidity under low-hanging clouds.

Finally, the theory was deduced that gas escaping from wells, pipe lines or tanks had collected in pockets, held down by the humid clouds and heavy atmosphere, until the automobiles, entering the gas pockets, had ignited the gas either from the engine spark or from a cigarette, and the fatal explosion had resulted. In the explosion that killed Witherup and Flanagan there was a tank only about sixty-five feet away from which thousands of feet of gas were escaping but which was not ignited by the explosion. That had occurred at a slight depression in the highway. It was generally believed that the gas, contrary to its natural tendency to rise and disappear, had been forced down by atmospheric pressure into this depression where it had been held until the automobile entered and exploded it. It was accepted merely as another freak of the already well-advertised Cushing field.

CHAPTER XVII

A New Era in Oil

THE CUSHING field came in on a rising market for crude oil. It was the first rising market of any importance which the industry had enjoyed since the Glenn Pool's output had flooded and depressed the market six years earlier. Wildcat promoters and drillers became more active than they had been for years.

Almost simultaneously with the rush to Cushing came a similar rush to Drumright, a few miles away. It was all in fact one great pool, so many miles in extent that it demanded more than one center of supply and trade. For a moment it appeared that there might be three. Drumright and Tiger identified themselves within sight of each other, but Drumright obtained the new postoffice and made the town.

Its mushroom growth in a way was even more spectacular than that of Cushing. Cushing was an established village when oil inflated it suddenly into a frantic frontier city. It had some slight facilities for taking care of the first rush of oil men. Drumright had nothing but a promise. Carl Blackman has told me that the first night he reached Drumright the only shelter he could find was a tent that offered sleeping space to eighty men, at one dollar each.

Every one of the eighty, he believes, stretched himself out for the night with a six-shooter under the makeshift pillow of his personal belongings. He smiles a little at that now. There was no need for all those six-guns. They were merely an indication of the spirit of the time and place. Those oil rushes were of a piece with the gold rush

to California in '49, the rush to the Comstock Lode in
Nevada in '61, the rushes to Colorado and Montana in
the '60's and '70's, the Klondike and Nome stampedes of
the '90's. They were high adventure upon frontiers
where the law of the six-gun was, for a time, the most
respected law of the land.

The mining rushes had drawn men from far greater
distances, some thousands of miles, but the lure of great
wealth to be quickly won attracted types fundamentally
the same. The newly opening Oklahoma oil fields were
closer to civilization, both in time and space, than were
the great mining developments of the preceding half century. Modernized small cities such as Tulsa and Oklahoma City were only a few hours away. Great commercial and industrial centers such as Kansas City, St. Louis
and Chicago could put men upon any piece of land within
the entire state of Oklahoma as quickly as a Tulsa resident
could have reached Muskogee a few years earlier.

But the days of the cowboy upon the lone prairie, the
bandit and the Indian, were still so fresh in the minds of
Oklahomans that a revolver seemed to be a logical part of
equipment. The fact that they were little used is beside
the point. They were part of the scene.

Stories which had helped to spread the news of vast
fortunes suddenly acquired in the oil fields had been
broadcast throughout the nation for a dozen years. Oklahoma produced the best and most popular of those stories
because of its Indian population and Indian ownership
of great areas of land which had proved to be incredibly
rich in oil. There is something especially interesting to
the average American in the story of an Indian whose ancestors had tortured and scalped the white pioneers and
in turn had been destroyed or exploited and restricted,
abruptly turning the laugh upon the exploiters. Perhaps it is the revelation of a delayed but poetic justice.
Perhaps it is our inherent sentimentalism, always respond-

ing to the Cinderella theme. Perhaps it is that recital of such examples of an abrupt smiling of Fate stimulates hope of some similar good fortune for us, regardless of our deserts.

The quirks of Fate, the folly of its beneficiaries, serve merely to make the stories more popular, more widely publicized. They stimulate our sense of superiority. They stir our resentment to ease our ego. They give us a small measure of satisfaction in the amusement and tolerant contempt with which we look upon the folly of ignorant individuals abruptly raised to a financial status above their capacity. If we had their luck we certainly would make far better use of it.

An old negro living in the Creek Nation in a one-room cabin with his wife and six children and nine dogs had eked out an existence for years on cornbread, sorghum and an occasional rabbit. He owned a forty-acre homestead, so barren that he could neither sell nor mortgage it. When a gusher came in on neighboring property he leased his land on a one-eighth royalty basis and collected a cash bonus that made him momentarily rich.

With his first amazing money he went to town. He bought a new calico dress for his wife, patent-leather shoes for each of his six children, and expensive collars for his nine dogs. Still he had money. He bought a shotgun for himself to improve the family menu with more rabbits. As a crowning touch, the extreme flight of his imagination, he purchased a piano. When the piano was delivered it was larger than the door of the cabin, so it was placed under a tree outside. It remained there for years as a roost for the birds and a toy for the children.

Another negro in similar circumstances indicated his capacity for high life and enjoyment of sudden affluence by buying himself a six-shooter, an organ, a silver-mounted saddle and a set of the most highly ornamented harness he could find. The organ, like the other man's piano, was too high to go through the cabin door straight up, although it would have passed if it had been turned down. The owner's i. q. was too low to comprehend that possibility. He simply sawed off the top of the organ and moved it in in pieces. Then he drove a spike into one side of the polished cabinet, hung the new saddle on it and placed the revolver conspicuously on top.

Wilsey Deer, a Creek woman whose sudden and un-expected fortune was invested in part in a ranch stocked with blooded cattle, butchered a $500 bull on the first occasion when she happened to need some beef. Another Creek used his first rush of wealth to buy a wooden horse as an appropriate ornament to his cabin yard, and pur-chased two phonographs to play two records.

No ordinary automobile would do for Lucinda Pitt-man, an Indian woman in the big money at Muskogee. Lucinda wanted color and plenty of it. Most of the auto-mobiles were black in those days. Lucinda picked out the gaudiest color she could find in a Muskogee store, pre-sented it to a Cadillac agent and demanded a car up-holstered and painted in that hue. The Cadillac com-pany demurred. After all, it was a dignified concern, and there was a limit. Could it afford to have a Cadillac pointed out with derision?

But cash was cash, and Lucinda had it. The company hesitated, and yielded. For a brief period Lucinda ap-peared daily upon the dusty roads around Muskogee, and as far away as Tulsa and Oklahoma City, in the gaudiest automobile in the United States. She made the Cadillac car the most popular automobile among the new-rich Indians. Cadillac profited beyond all expectations. Before long, pink and crimson, baby-blue and sea-green cars were a new vogue throughout the United States.

The story of Jackson Barnett, previously mentioned in these pages, is still alive in the news from time to time. Barnett was an "incompetent," legally and in fact. His

vast fortune was administered under the control of a
guardian. His pleasures were simple and childlike. His
chief interest was in a few beloved horses. His imagination was incapable of such publicity-winning flights as
flame-colored Rolls-Royces or registered bulls butchered
to make an Indian feast. But there were plenty of persons who had an eye upon his riches and were eager to
do what they believed to be the right thing with such
wealth. Among them, by marriage and otherwise, they
managed to spread the money and the fame of easy oil
royalties far and wide.

Other Indians, not so wealthy but wealthy enough,
managed to do pretty well in their own right, without the
help of guardians or wives. One of the new-rich Indians in Muskogee, for example, having had his curiosity
aroused by tales of the great stockyards in Chicago, took
a taxicab for the 700-mile trip to see them. For a moment it gave the Indian and his oil fortune publicity
comparable to that attained by Death Valley Scotty on
an earlier journey from Los Angeles to Chicago.

One young Osage woman, appearing at a public recep-
tion in the White House, attracted attention because of
the spectacular beauty and extraordinary grace of her
costume. When a society reporter covering the event for
a Washington newspaper needed an unusual sidelight for
her story she sought to gain the Indian girl's confidence
by complimenting her on the beauty of her gown. Shyly,
modestly and haltingly the dusky maiden explained that
she had not made the dress herself. She had purchased
it in Paris. That was news.

Eastman Richards, a Creek whose barren acres had
been assigned to him by the Dawes Commission in the
belief that if enough corn and rabbits could be produced thereon to sustain life another Indian problem
would be settled, found himself so wealthy that nothing
less than an entire town could serve as a fitting monu-

ment for his eleemosynary imagination. Richards, un-
deterred by guardianship, selected a site in McIntosh
county and built the town complete as a home and haven
for himself and his less affluent fellows. Richardsville
cost a million dollars. It doesn't look it today, but it was a
million dollars' worth of publicity.

With similar stories finding their way into the newspapers from Muskogee to Pittsburgh and from Tulsa to
Los Angeles, the Oklahoma oil fields became the goal of
countless adventurers. Earlier arrivals, naturally, had
been men with a background in the oil development of
Pennsylvania and elsewhere. They were men "born in
the shadow of an oil derrick," as the saying went. They
were practical men like Tom Slick, who could dif-
ferentiate between the odor and texture of Bartlesville
sand, Wheeler sand, Wilcox sand and the various other
sands which proved to be oil producers. There were
drillers like Guffey and Galey, W. H. Heydrick, Amos
Steelsmith, Cam Bloom and innumerable others—men
who actually knew how to build a derrick and sink a well.
There were experienced workmen in all phases of the
actual task of oil production. There were men of industrial and financial acumen such as H. V. Foster, Pat White,
Harry Sinclair, C. B. Shaffer and the Phillips brothers.
There were men of energy, daring and promotional ability such as Josh Cosden.

The oil was there, and all the machinery, human and
mechanical, necessary to its production was available.
Refineries and transportation facilities were established
and improved just as the potential demand for the product
was improved with the growing automobile industry.
Gas by the hundreds of millions of feet continued to run
to waste, but the actual running of oil into creek beds and
makeshift earthen reservoirs was reduced.

For example, the McMan Oil Company, organized by
McFarlin and Chapman in the early days of the Cushing-

Drumright development, built a new steel tank of 55,000-barrels capacity every day for 114 consecutive days, and filled them with McMan oil as fast as they could be constructed. With similar provisions made in various areas even a fluctuation ranging between thirty-five cents and three dollars and a half per barrel in the price of oil, which actually took place in a period of thirteen years, could not stop the oil men. But it did tend to bring about reorganization of the business in general, to take it out of the hands of the smaller independents who led the earlier development and bring it under the control of larger corporations.

Of the three concerns of national significance in the earlier field, Prairie Oil & Gas, as a recognized representative of Standard Oil, had been the first most powerful, although Indian Territory Illuminating Oil Company with its rich holdings in the Osage country was a close contender. The Texas Company, in part because of its refineries and pipe line developments, and the Gulf Oil Company, controlled by the Mellon interests in Pittsburgh, were close in line.

But the tremendous production of the Cushing-Drumright area brought such concerns as Shell, Mid-Continent, Carter, Pure Oil and others into the picture in an important way. The Skelly Oil Company had its inception there, although it was not to be incorporated as such until a few years later. The Magnolia Petroleum Company, Ohio Oil Company, Stanolind Oil & Gas Company—all associated with the Standard—the Continental, Phillips, Empire, Barnsdall, Tidewater and others came in with a rush.

The old-time individualists who preferred to sell properties rather than to develop, produce and market the oil itself could find buyers among these great corporations. McFarlin and Chapman, for example, only five years after their incorporation of the McMan Oil Company with a capital of $100,000 for operation in the Cushing-Drum-right field, sold to the Magnolia Petroleum Company, presumably representing Standard Oil, for $35,000,000. Waite Phillips found a $25,000,000 cash customer in Blair & Company, presumably acting for Barnsdall, under obligations to Standard. Such gigantic deals, of course, were spread over a period of time, through the war years and the further development of the automobile industry, but their inception was in the new era which began approximately with the proving of the Cushing-Drum-right field.

A new type of oil men, brokers, bankers and lease-buyers and salesmen began to make themselves felt in that new era. Most notable, as he has been most widely publicized of the group, probably is Harry Sinclair. Sinclair had not hesitated at the very start of Oklahoma's oil production to put his small cash resources into the business. But as the business and potential profits expanded faster than his resources he realized the possibility of multiplication of profits by judicious use of credit. Sinclair became the first great exponent of the use of "O. P. M." in Oklahoma's oil business. By that was meant "other people's money." It was other people's money directed not only to the multiplication of Sinclair's profits but the profits of the lender or investor.

How well it worked is revealed in the history of the Exchange National Bank of Tulsa, which was the first clearly defined medium of its use, organized primarily as the oil men's bank. Its growth, simultaneous with the spectacular and substantial growth of the city of Tulsa itself, is a revelation. That dramatic business history has been written by Mr. Harold Mueller in a series of articles published by *The Oklahoma City Times*. In the first six years of its history the Exchange National, organized at the instigation of Sinclair and White, as has been explained, to give the oil men service after the failure of the

Farmers National, increased its resources from $410,000 to $11,000,000.

In the same period Tulsa itself, in part supported by the bank and in part supporting the bank, was making itself indispensable to the oil business of Oklahoma. Oil companies were encouraged in every possible way to make headquarters there. Urgent invitations, supported by able arguments, and at times by the free gift of building sites or bonuses, were sent to oil-equipment companies, tool companies, lumber companies, machine shops and every other type of business that had anything to do with oil.

Office buildings were erected primarily to house oil companies or related concerns. Hotel accommodations were steadily improved. Communications were improved, most notably with a great bridge across the Arkansas River. Refineries were established.

Oil prices continued low and further prospecting and wildcatting was comparatively slight in 1910, the year of organization of the Exchange National. Output of the famous Glenn Pool was declining. Despite the opening of a few small pools, the total production of the state for the year and the total cash receipts for crude were only slightly higher than in the preceding year. Completion of the Oklahoma Pipe Line to Baton Rouge helped a little, and gave promise of more assistance for subsequent years.

By the time the price had risen to one dollar a barrel and production doubled, in 1913, all the machinery for profit-making and oil and business development was substantially organized. An important part of the machinery was the oil bank. Harry Sinclair made a fortune in the Cushing field, and the Exchange National was piling up assets and deposits in Tulsa. In its first year it absorbed the Union Trust Company; then the Colonial Trust Company.

When the vast production of the Cushing Pool was

supplemented by the spectacular output of the newly discovered Healdton Pool and almost a score of other smaller
fields prices fell away again in 1914.

But war was declared in Europe. Before the year was
over a new demand for gasoline, lubricants and fuel oil,
such as the world had never known and even the most optimistic oil men had never imagined, began to make itself felt. Prices began to rise again. In the second year of
the war the Cushing Pool increased to a peak production
around three hundred thousand barrels of oil a day.
Independent prospectors or wildcatters opened more than
a dozen smaller pools in eastern Oklahoma. The following year, with Cushing slowly declining, but prices rising
steadily, more than twenty small new fields were brought
into production.

Harry Sinclair began to see for himself a place in the
sun in which he could hope to compete even with such
fabulous figures as the Rockefellers. He had helped to
make Tulsa a fine city, but it was still not great enough
for his soaring imagination and ambition. He moved his
headquarters to New York and organized the Sinclair
Crude Oil Purchasing Company to be a concern of international scope and power.

Patrick White resigned the presidency of the Exchange National Bank, and Harry Sinclair's brother, E.
W. Sinclair, was advanced to the office. The bank had far
outgrown its quarters. It proceeded to build a new home
worthy of itself and of the city with which it had arisen
to wealth and power. A modern twelve-story building
was erected at Third Street and Boston Avenue, in the
heart of Tulsa.

Tulsa expanded with pride and wealth. In the year
between its decision to build and the opening of the new
building on Nov. 11, 1917, the resources of the bank almost doubled, increasing from $11,000,000 to $21,000,-
000. That was a measure of Tulsa's growth and the

growth of Oklahoma's oil-given prosperity. The Exchange National absorbed the Planters National.

A trust department which would be competent to take over, invest and protect some of the millions of dollars which were being derived from oil but were going out of the state into trust companies in New York and elsewhere seemed to be the next logical step. Oklahoma business should be diversified, and Oklahoma money invested in such diversification. E. W. Sinclair, president; Robert McFarlin, chairman of the board, and H. L. Standeven, experienced in estate matters as a probate judge, made some investigation. The Exchange Trust Company was organized as a subsidiary of the Exchange National Bank. Millions of money which otherwise would have gone to the East for investment were kept in Oklahoma to the profit of the Exchange National, the state and the individuals interested.

And then came the post-war depression of 1920. In a situation tense with danger and a threat of panic, the American National Bank, which had been a minor rival of the Exchange for some years, found itself in difficulties. A failure would be disastrous. Every banker and important business man in Oklahoma who knew the situation recognized its gravity. The closing of the Farmers National ten years earlier, when Tulsa was a town of 15,000, had threatened a major financial catastrophe until the organization of the Exchange National had relieved the situation. And in 1920, with a town of 72,000, and many more millions of dollars and many more hundreds of depositors and businesses of greater variety endangered, no one could predict the outcome.

The Tulsa Clearing House Association conferred all day and all night. There seemed no practical, businesslike way of saving the American Bank. The exhausted conferees went home at daybreak, hoping and praying that the expected disaster would not be quite so terrific as they feared. Robert McFarlin, continuing to ponder every phase of the problem, could not sleep. At eight o'clock he was back in his office looking across the street to the doors of the American National.

Already there was a little knot of worried citizens at the steps of the bank. McFarlin knew many of them by sight. Some were his friends. Every one, he suspected, faced disaster with the loss of all his immediate cash resources. To the individual with all his money in that bank, whether it was one hundred or ten thousand dollars, the danger was equally serious. Each person in the milling growing crowd seemed the center of an impending tragedy.

McFarlin watched the crowd grow in numbers and excitement. And at nine o'clock, when the doors of the Exchange National opened as usual while the doors of the American National remained closed, the voice of the crowd lifted in one united moan. The shade upon the closed door ran up, and a man inside pasted upon the glass a formal notice: "This Bank Closed by Order of the Controller of the Currency."

Threatening voices punctured the groaning murmur of the crowd that filled the street. McFarlin turned from the scene and hastily called another meeting of his directors. The directors' room looked out upon the scene of the drama. No more argument was needed. When McFarlin repeated a suggestion which had been made and rejected in the long conference of the previous night—that the Exchange National should guarantee the deposits of the American National, and find what resources it might in the wreckage—it was approved at once.

Many of those bankers had known poverty. All had known the need of funds available immediately to take advantage of opportunity. They realized a responsibility to their profession and to their community. Without waiting to work out details of the plan they approved it

in principle, and without a moment's delay Robert Mc-
Farlin hurried to the street, shouldered his way through
the muttering crowd and vaulted to a point of vantage
upon a parked automobile.

As he lifted his hand for attention there was a threat
in the hoarse voice of the crowd. Here was another member of the fraternity which had taken their money and
wasted or lost it and left them to suffer. McFarlin ignored
the menace.

"Quiet!" he shouted. "Quiet, please!"

The mutter of the crowd and the shouts and threats
of a few most excited persons continued.

"Quiet!" There was authority in that voice. The
confused noises of the crowd subsided. "I am chairman
of the board of directors of the bank across the street. I
am Robert McFarlin. Some of you know me. You know
my word is good."

A murmur of assent arose in the crowd. Here was hope.

"You will notice that bank is open for business," Mc-
Farlin continued. "You have my word that it will be open
tomorrow, and that it will honor your checks and pass-
books to the full extent, one hundred cents on the dollar.
You have nothing to worry about. Come to the Exchange National tomorrow and get your money, if you
want it."

He leaped down from the automobile and made his
way back to his office. The mob melted away. And
promptly at nine o'clock next morning the Exchange
National tellers began paying off the American National
depositors. Before the morning passed some of the first
withdrawals were redeposited in new accounts with the
Exchange National. Before noon the run was over. Many
did not even make an effort to withdraw.

Later unofficial reports asserted that that action cost
the Exchange National more than $100,000. But the
Exchange had the money for the purpose. A local panic

was averted. Suffering of numerous individual depositors was prevented. Business was strengthened and stabilized. The bank itself gained important new customers and business.

Within two years after that event, five years after it had moved into the new banking building which had been planned to be big enough for all time, the Exchange National found itself so crowded that larger quarters were required. A twelve-story addition was added on Boston Avenue, and occupied in the following year. In the six-year period the institution's resources had risen from $21,000,000 to $33,000,000.

E. W. Sinclair had resigned the presidency and gone to New York to join his brother, as president of the Sinclair Consolidated Oil Corporation. Robert P. Brewer, a native Oklahoman, with wide banking experience, had been made president and then chairman of the board of the Exchange. James J. McGraw, risen from a grocery store in Ponca City, through oil, banking and war service which won him the Cross of the Legion of Honor, became the new president of the Exchange.

McGraw enjoyed important and powerful political connections and a wide acquaintance with European interests. The Exchange National had already become known as "the Oil Bank of America, in the Oil Capital of the World." It enlarged that reputation under the new president. The bank had customers and depositors active in virtually every major oil field on earth. Russia, Roumania, South America, Mexico dealt with the Exchange National. Pay checks of operators, superintendents, geologists, drillers and tool dressers found their way from all quarters of the globe to the Exchange National for deposit.

Under those conditions the bank and its affiliated trust company boomed as the oil business and Tulsa boomed through the years of sensational prosperity in the Harding and Coolidge administrations. While the steel frame of a third addition mounted to a height of twenty-eight stories above Tulsa's streets President McGraw was stricken with an acute illness which caused his death. He had been an important figure not only in the development of Oklahoma but in the economic and political life of the nation. Telegrams of condolence from such a variety of men as President Coolidge, Chief Justice Taft, John McCormack and Irvin S. Cobb indicated the variety of his accomplishments and the power of his personality.

But Oklahoma had developed many powerful men, as it had been developed by them. Harry H. Rogers, with long and successful experience in the law, the oil business and banking, was one who had an interest in the Exchange National. As international president of Rotary he had traveled far and wide. He had been signally honored by King Albert of Belgium for his work in the interest of international peace. He was a logical choice for the presidency of a bank which had become in effect an international bank in its relation to the oil business. Under him in the great days of 1928 and the spring of 1929 expansion further accelerated.

By the time its final monumental home was occupied in October, 1928, the combined resources of the Exchange Bank and its affiliates had passed the $60,000,000 mark, a growth of $59,590,000 in eighteen years. In the following year it bought control of J. E. Crosbie's rich and substantial Central National Bank and Trust Company. Deposits passed the $50,000,000 mark. New capitalization was indicated, and promptly arranged. Stockholders increased from 184 to 450. Combined deposits passed $60,000,000. Combined resources passed $110,-000,000.

The bank's trust and investment affiliates financed the Wright building, the Tulsa Hotel, the Bliss Hotel, the Alvin Hotel and numerous large apartment buildings in

Tulsa, and important hotels and other structures in other cities of the state.

Then came the market crash of 1929.

Values melted. Oil prices slipped. Within a few weeks call deposits decreased by more than $4,000,000. The stock market continued to decline. In two years the price of oil in Oklahoma dropped to twenty-nine cents a barrel. It had been $3.50 in 1920. The great bank's assets froze. Men who had been fabulously rich, leaders in the oil industry, were known to owe the bank great sums of money which they could not pay. Rumor and threat of panic appeared. Depositors, hard pressed or timorous, continued to withdraw their money. J. A. Chapman, erstwhile partner of Robert McFarlin in the McMan company which had pumped an original well into a business which brought them $35,000,000 from the Magnolia Oil Company, hurried into Tulsa to investigate the perilous situation.

It would require ten million dollars to forestall disaster, he decided. "I'm willing to put up $5,000,000 if the other fellows will put up another $5,000,000," he told his associates. Chapman had had as much as $1,-000,000 on deposit in the bank at one time and had never borrowed a dime from it. But the fortunes of all the others were not so liquid. The rescue plan was not worked out. But some of the more doubtful frozen assets were segregated and taken over by the men who had the means to do it. Chapman himself took $700,000 worth of the paper. Rogers' health failed under the strain and he resigned.

Chapman, McFarlin and the Sinclairs decided on desperate measures to save the institution, the depositors and themselves. Elmore F. Higgins, vice-president of the National City Bank of New York, was brought in as president to succeed Rogers. His cold, impersonal, Wall Street practices were not popular with the Oklahomans,

reared in a tradition of more friendly business relationships. Frozen mortgages were arbitrarily assigned to investing clients. It was sound business practice in the
emergency but the clients frequently resented it. More
cash was required.

In the course of a few months in 1932 Chapman advanced more than $4,000,000. It was not enough. The
great bank seemed doomed, and with it nearly one hundred smaller Oklahoma banks which had reserves on deposit there.

No one in the United States who had reached years
of normal adult intelligence, or who had a single dollar
in any bank in the country in the spring of 1933, will
forget that period. Governors of twelve states had already
declared banking holidays in the hope of giving financial
institutions an opportunity to adjust affairs before all
failed completely. With the catastrophe of failure of the
Exchange National threatening, Oklahoma became the
thirteenth.

On March 1, President Higgins notified Washington
and all correspondent banks that the Exchange National
could not open on March 2. But to save it from confessed failure which would immediately wreck the entire
banking organization of the state, Governor Murray declared the holiday. The national bank holiday followed
promptly. Disaster was forestalled for a time. But when
that shadow was lifted the troubles of the Exchange National were far from finished.

Very soon it became apparent that the institution could
not survive as it had grown. In May, Chapman and others
contributed $200,000 more in cash. It was useless. The
great and famous Exchange National Bank closed its doors
for the last time. When it reopened the following day,
reorganized with $4,000,000 of Reconstruction Finance
Corporation funds, $1,750,000 contributed by the faith-
ful Chapman, $2,125,000 from the Sinclair interests, and

$125,000 from such members of the board as had a few dollars of their earlier millions left, it was the National Bank of Tulsa. It is so today, though half the residents of the city of Tulsa still refer to it as the Exchange National.

The curtain had fallen on one of the most exciting acts of spectacular success and spectacular failure in the great drama of Oil on the Last Frontier. But that was not all.

The state of Oklahoma, the oil capital of Tulsa, and oil men and companies around the world heaved a sigh of relief. Business could go on. Life could go on. And then, more abruptly than the panic itself, came news that twenty-eight former directors of the Exchange Trust Company, including many of the wealthiest, most powerful and most widely respected citizens of the state, must face charges of embezzlement. It was stunning, inconceivable.

J. A. Chapman, who had contributed many hundreds of thousands of dollars from his personal fortune to preserve the tottering bank, a thief? Harry Sinclair, a thief? H. V. Foster, a thief? Twenty-five other leading citizens, pillars of society, philanthropists, churchmen, economists, builders of their communities and their state, all thieves? It was impossible. But the charges were filed, the defendants named, the evidence against them assembled.

Specifically, twenty-five of the erstwhile directors were charged with the embezzlement of $1,000 in March, 1931; $6,650 in April; $1,600 in September, from the estate of George L. Miller, of the famous 101 Ranch near Ponca City. That appeared to be sufficiently definite to be damaging, but the fact that the charges were based entirely upon technicalities of the trust department's records brought about a wave of public resentment. The whole action was declared to be persecution rather than prosecution. It intensified rather than suppressed the drama.

When the preliminary hearing began before Judge

Bradford Williams of the Common Pleas court in Tulsa the Judge brought three associates to the bench with him, explaining that, while the rulings and decisions would be his, his fellow judges would be there in an advisory capacity. The courtroom was crowded with legal talent and with leaders in almost every phase of Tulsa's and Oklahoma's business life. Four bailiffs were required to handle the crowd. Twenty-seven lawyers represented the defendants. Seven lawyers appeared for the state to formulate the prosecution. Four court reporters worked in ten-minute shifts. It was what Hollywood might describe as a stupendous super-colossal trial spectacle.

We need not go into details of evidence here. With the opening statements of prosecution and defense completed, the trial lapsed into such dry detail of testimony on the technicalities of trust-fund bookkeeping that spectators and principals alike were bored to tears. It went on for days. The drama collapsed. In the end the criminal charges were dismissed. The Titans of Tulsa were officially cleared.

But they were not elated. They walked sadly from the courtroom. Their pride had suffered a blow from which it seemed unlikely to recover. But business must go on.

CHAPTER XVIII

SEMINOLE'S HELL AND HEAVEN

IN AN EFFORT to complete the picture of the rise and fall of the Exchange National Bank as an illuminating sidelight on the drama of oil in Oklahoma, I have run ahead of the proper chronological sequence of the narrative as a whole. The main thread of the story should be picked up where it was temporarily laid aside with the development of the Cushing field in a preceding chapter. Even so, all the intervening incidents cannot be given space.

There are in Oklahoma today some hundreds of more or less independent oil pools. Many have produced and many are still producing great riches. Innumerable pools in the Osage country, around Bartlesville, around Muskogee, gave the state, the Indians and various individuals their first taste of power and wealth. The names of perhaps a dozen pools not yet mentioned in these pages have impressed themselves upon my own memory in the course of interviews with oil men. The Hogshooter field, Boston, Burbank, Nowata, Tonkawa and many others deserve more space than they can be accorded. They have meant life and death, fortune and failure, poverty and power, to thousands of persons. Statistics of production, the rise of towns based upon their output, a thousand individual dramas within them, would thrill oil men and associated businessmen, but they cannot be given here. There were too many. Highlights must suffice to reveal the widespread development.

Of these highlights the Glenn Pool was first, and the Cushing-Drumright development the second great event. It was there that many of the fortunes which became the constructive force of Oklahoma's advancement actually became great. The Indian lands were the first rich oil lands. But the movement of the wildcatters, the true prospectors and pioneers of oil development, extended steadily throughout the northern and eastern and gradually into the southern and central parts of the state.

So in due time, after some hundreds of smaller pools had been discovered and exploited came the sensational riches of the greater Seminole field. The Cushing field had passed its peak production of 300,000 barrels a day in 1915. In the following year the Healdton area reached its peak of 90,000 barrels a day. By the end of the war Cushing's output had decreased so much that the state's total production would have been jeopardized if the Burbank field in Osage County and the Tonkawa Pool on the opposite side of Ponca City, among others, had not been coming into importance.

Burbank at first was looked upon with little favor. The first leases there were granted on a bonus of only ten dollars an acre. But the speed and riches of the output are revealed effectively by the fact that within a year leases were selling for as much as $10,000 an acre. For a moment the eyes of all the leading oil men of the state and nation were upon Burbank and Tonkawa. But then came Seminole. That was a heaven and hell of oil production and all the wild activities attendant thereon.

O. D. Strother, originally a shoe salesman whose traveling had made him intimately acquainted with the land and its people, was the man of faith and vision behind the Seminole venture. For years Strother had made himself almost a nuisance among his friends in Tulsa and elsewhere because of his insistence that there was oil in the Seminole area. Every dollar he could obtain went into the purchase of Seminole lands and leases. As time went on without a well proving the validity of his belief the problem of paying taxes complicated the problem of acquiring more acreage.

The mud hazard on Seminole's main street in the early days of its oil boom. The condition was typical of numerous oil-made towns in the days before streets were paved.

But Strother's faith never wavered. Friends who honored him as a pioneer were impressed by the certainty of his belief. He was the prophet of oil in a barren land. But few believed. A man with the tools, skill and money to justify his faith was needed. Such a one appeared in the person of Robert Garland, operating for the Independent Oil & Gas Company on a lease a mile east of the village of Seminole. Garland had entered the oil business as a youth of twenty-two, a laborer in the initial development of the Cushing field, thirteen years earlier.

He had advanced from a job as roustabout to the dignity and skill of a driller. By competence in his job he had gradually equipped himself to extend his opportunities. The Tulsa *World,* summarizing his activities after success had made him famous, asserted that he had a wonderful record of dry holes. But he was undiscouraged. "Fill the earth full of enough holes and you'll find it some time." That was his plan of operation.

Early in his career he had seen some examples of the element of luck in the Cushing field. He had seen or heard of the loss of a string of tools in a well owned by Frank A. Gillespie raising production of the well from 4,000 barrels to 18,000 barrels. Ordinarily the loss of a string of tools in the course of a drilling operation is a serious inconvenience if not a disaster necessitating the abandonment of the hole and the loss of all the money put into it. But in the well involved in this accident nature and luck combined in almost unbelievable fashion to turn threatened disaster into triumph.

The well had come in with a production of 4,000 barrels when the tools were lost. Fishing operations were started promptly in the fear that the tools might block the flow of oil, and necessitate abandonment of a rich well which had cost many thousands of dollars. But when the fishing mechanism was lowered in an effort to pick up the lost tools, it encountered a phenomenon that gave the workmen a shock. From deep within the earth came a blow upon the fishing tool. It was a blow of metal against metal, repeated again and again.

Old-timers among the workmen recalled a Paul Bunyan-like story attributed to Pete Purchill by Harry Botsford, writing legends of petroleum in the *Oil Trade Journal* some time earlier. The imaginative Purchill had made a fantastic tale of tools bouncing on rubber rock 1,100 feet deep in Amos Hodge's Tickle Top well. In Cushing the legend seemed to have become reality. It was not long before geologists and experienced drillers explained the phenomenon more prosaically as due to gas pressure which lifted and dropped the tools at brief intervals, somewhat as steam generated in the depths of a geyser lifts the weight of water above it.

But by the time that simple explanation had been registered the bouncing tools in the Gillespie well had opened the hole into richer oil sands which raised production from 4,000 to 18,000 barrels. That was luck which helped Gillespie on the way to a fortune which made it possible for him to promote a private reclamation scheme at Gila Bend as one incident of his career. In that scheme he is reported to have expended $3,000,000 only to have the whole project thwarted when an adverse turn of luck filled his vast reservoir with sand.

Perhaps a word of explanation should be interpolated for the lay reader who might wonder why a well which has struck oil should not always and deliberately be drilled deeper to obtain more oil. Geologists and practical drillers have found that oil lies in beds of various sands at various depths. Above or below these beds of oil-bearing sand there are strata of rock and clay, and occasionally areas impregnated with salt water. The trick is to tap the oil sands at their upper levels, but not to drill so deeply that the well might extend into salt water until as much oil as possible has been drawn off.

An interesting incident illustrating the care with which that problem is approached, and a bit of phenomenal luck similar to that of the bouncing tools, has been told to me by John Winemiller, a veteran driller still operating from offices in Tulsa.

George Simon, a well-shooter who had been employed in Egypt by a British syndicate, interested his principals in the Oklahoma fields and came to the rich Beggs Pool at about the time of its opening. He drilled a number of wells and finally came to a parting of the way with his employers. In final cancellation of his contract he was given a clear title to an insignificant little ten-barrel well near Beggs. Ten barrels was hardly worth pumping. Simon, fearful of sinking entirely through the sand but desperate in his need of improving the well to point where he could sell it at some profit, decided to sink another foot. Cautiously, fearfully, he did so. There was no improvement, but no salt water. He tried another foot with the same results. He would go one more foot, if he busted.

And at the end of that third foot there came a rush of gas-driven oil which flowed 6,000 barrels in a day. Simon took 4,000,000 barrels of oil from that development, cashed in for $4,500,000, and his name was added to the roll of Oklahoma oil fields.

Similar incidents of luck, and the fantastic rewards of persistence supplemented by luck, were familiar to Bob Garland when he sunk his first well in the Seminole area. He knew he would strike it sometime. He did. The fact that O. D. Strother, the persistent prophet of oil in that region, did not live to see his faith justified beyond his wildest dreams was a tragedy quickly forgotten by the thousands who battled and labored for the wealth of Seminole. And how it flowed!

"All the majesty, the fury, the danger and the urgency

of a great oil boom is at Seminole," a reporter for the
Tulsa *World* was inspired to write. "All its confusion and
strife and litter; its laughter, its suffering, its curses and its
victories. And all of its raw material. . . .

"It has taken skill, patience, perseverance, the over-
coming of almost unsurmountable obstacles, the hardest
of manual labor, and millions of dollars to bring in the
great Seminole field. And in pride, perspiration and tears
the oil fraternity has done it."

Realization of a dream of a field of gushers with 527,-
000 barrels of oil flowing in a single day—July 20, 1927—
was more than even the oil fraternity had bargained for.
The market broke under that tremendous pressure. The
world had too much oil. Operators of the mid-continent
field faced a serious economic problem. They appeared
to be wasting their own resources and jeopardizing their
chances for continued profits from oil by releasing those
reservoirs for waste and deterioration. They met and
elected Ray M. Collins to attempt the colossal task of restricting and prorating production so that oil which came
forth might be sold at some profit and the remainder conserved below ground as it had been preserved by nature
for millions of years.

It was a difficult task. Had the larger companies been
alone in the field with the intelligence and resources to
withhold production to a point at which supply could be
regulated to demand it would have been comparatively
simple. But there were hundreds of independent lease-
owners who were forced to drill to recover the money
invested in leases and to gain their profits. Restriction
threatened to break and destroy them.

The various original small pools in the field, including
Searight, Earlsboro, Bowlegs, Little River and Seminole
proper, were from one to several miles apart, but as intervening acreage was leased and tapped by independent
wells they began to merge into a single great pool. The

cost of leases on intervening and neighboring ground reached staggering heights. In those days of so-called Coolidge prosperity there was plenty of speculative money available to buy.

The fact that an average of $65,000 was required to drill a single well, wet or dry, in that difficult geological formation, was no deterrent. But such investments arranged and applied by independent promoters called for cash dividends if the work of applying "other people's money" to oil development was to go on. The geological formation required special drilling rigs and a change of rigs at a depth of 3,500 feet. The fine sand driven by tremendous gas pressure destroyed parts of the equipment and casing as an emery wheel might do. When tools were lost the fishing for their recovery was especially difficult. Frequently a rig had to be skidded seventy-five feet from its first location and a new hole started. The oil men estimated that it cost as much to drill one hole in the Seminole field in 1927 as to drill twenty average holes in Oklahoma fifteen years earlier. But still the oil poured out, the money poured in and the drilling continued.

Heavy rains upon the unpaved clay roads of the area made transportation a problem which added to costs and to confusion. The area was literally swamped in oil and mud. Perhaps no one who was not upon the scene, unless it might have been the soldiers and service of supply men upon the battlefields of Europe in the winters of the World War, could imagine the depths and dangers and horrors of that mud.

With hundreds of passenger cars carrying speculators, lease-buyers, workmen and superintendents into the field, the churned depths of fluid mud upon a base of sticky mud became a revealing test of the fortitude of man and the power of the automobile. Hub-deep, running-board deep, fender-top deep, they called it with wry grins as they slipped, skidded and floundered on. When it became windshield-deep they left the cars and struggled on afoot or horseback—if they could find a horse.

"On all the roads and trails there are heavily loaded trucks that wallow and roar imperiously but many times helplessly in great sinkholes of the sticky black mud," wrote a reporter for the *World*. "Or else some lighter car, not yet acquainted with the numerous high centers of the roads, finds itself perched precariously but permanently upon some such road curse with its two wheels spinning playfully in the water around it like a child in wading."

Traffic jams were terrific. Bridges and culverts without guard rails, with skidding approaches, upgrade and down, turned hair grey and shortened lives. Consider the verbatim report of one of the men who was there, C. S. Choinski, field clerk of the Empire Oil and Refining Company, as written for the tenth anniversary edition of the Seminole *Producer*.

"In December, 1926, I received orders at Cushing to go to Seminole," Choinski wrote. "I had heard enough from our truck drivers to know that the Shawnee road was impassable, and so decided to come into Seminole by way of Okemah, Holdenville, Wewoka and Bowlegs. Two miles east of Bowlegs I found the first bad road, but having a Model T, I was better able to navigate than others.

"Bowlegs presented a motley sight, with lumber scattered everywhere, everybody rushing around and hammering and sawing on a hotel building, stores, dance halls and movie houses. Most of the people lived in shacks and tents.

"Leaving north, I found the road from there to Seminole simply terrible. (I have always wished I had seen the Shawnee road.) Ruts hub deep. One continuous line of cars and trucks going and coming. No chance to drive around a slow moving car. And they were all slow moving, because if one got stuck it held up traffic for a mile or more. You could detour around several especially bad mudholes by paying the owner of adjacent land fifty cents or a dollar. The owner was usually around with a shotgun. But when you came to the end of a detour you might have to wait for two hours to get back into the solid line of cars.

"The entire road was lined with small shacks and tents, some doing business. . . . This trip required a total of fourteen hours to negotiate the five miles from Bowlegs to Seminole. Today it requires less than five minutes.

"In Seminole it was the same. Mud everywhere. Many cars were stuck in the mud on Main Street. What sidewalks the town had were piled high with boxes of goods. The Chamber of Commerce building was a two-by-four shack built in the street.

"Truck drivers had to wait in line for a day or two to get to the freight depot."

But the oil was more extensive than the mud, as it was deeper—some thousands of feet deeper. Seminole had everything that the most colorful frontier towns and most spectacular mining camps in the history of the United States had revealed in three-quarters of a century preceding. To be sure, its saloons were not the open-door palaces or dives which have come down in history of mining camps from the time of Flood and O'Brien's famous Exchange in the early days of gold-mad San Francisco, the Crystal bar in Virginia City, Nevada, in the days of the big bonanza, or the hell holes of Alder Gulch, Montana. Prohibition made the liquor business a bit more furtive, but bootleggers experienced in their trade through practice against the Federal amendment were able to supply the demand with a minimum of difficulty. The dance halls and red-light district centering around notorious Bishop's Alley in the North End were as full of evil as hell itself.

An indication of the character of the dance halls may be found in a prominent sign posted in the North End Club. "No Firearms or Knives Aloud Here." That sign was needed, although there is some indication that it was not effectively enforced. In the first ten years of Seminole's sensational life as an oil metropolis there were thirty-three arrests for murder. And every man arrested by the local police on that charge was convicted. In the same period seventeen were convicted of manslaughter, and eleven convicted of assault with intent to kill. Seminole was a tough town, but in the main the law was well enforced. There was a substantial backing of honest, well-to-do, public-spirited citizenry.

Perhaps the most notorious criminal ever associated with the vicious lot in Seminole's North End was "Pretty Boy" Floyd. Floyd was employed in the neighborhood of Seminole in the early boom days, but was never convicted of crime there. One of his associates, George Birdwell, was killed in an attempted bank robbery.

"Tangle-Eye" Hall, who operated the Big C dance hall for a time, went to the Federal penitentiary at Leavenworth. Delmar Hull and "Dutch" Pugh died with their boots on in a hi-jacking enterprise of prohibition days. "Baldy" Burton went to the Missouri state penitentiary for a fifty-year term. Adam Richetti, an associate of "Pretty Boy" Floyd, was convicted of a part in the Kansas City Union Station massacre, and sentenced to death. White Walker, who once terrorized the neighboring town of Bowlegs for an entire night and was generally looked upon as the toughest man in the area, was shot to death by officers in Texas. Hub Cooper, serving time for bank robbery, died in the Arkansas penitentiary.

So it went. Not all the bad actors paid the death penalty or even served prison terms. A few worked their way into legitimate business or trades. Some of the more notorious women reformed, married and settled down. Some settled down without reforming or marrying. After the flush days of the Seminole boom Opal Cooper established a bawdy house in a Texas oil town. Barrel House Sue moved to a northern city. Big Emma Smythe moved to the Texas fields.

As tough or tougher than Seminole in the first mad boom was the neighboring town of Bowlegs which came into the big money, the big excitement and the big publicity at the same time. Before the discovery of oil the settlement consisted of little more than a general store and half a dozen houses. It had neither railroad nor post-office. Its name, said to have been derived from a Seminole chief known as Billy Bowlegs who with his entire family had been killed in an Indian feud in the neighborhood, was its chief claim to distinction.

But with the opening of the first rich oil well in the neighborhood the tiny hamlet found itself flooded almost overnight not only with oil but with speculators, promoters, lease-buyers, oil-workers, storekeepers, dance-hall girls, café operators, swindlers, gunmen and toughs of all descriptions. The mad life of a frontier boom town coupled with the unique name which it bore made it exceptionally picturesque copy for metropolitan newspaper correspondents. The names of its most notorious characters such as Big Nell, the bootleg queen, Spanish Blacky and Wingy McDaniel were equally in keeping with the color of the town's night life. It wasn't long before Bowlegs became widely publicized as "the toughest town in the United States."

There was some justification of that reputation. The first flimsy calaboose could not begin to hold the mobs of drunks, crooks and toughs thrown into it nightly. They literally burst the walls and escaped to continue their activities, sometimes in the Bowlegs dives from which they had been dragged and sometimes in neighboring Seminole. So many escaped that the jail had to be rebuilt time after time. There was so little interest in law enforcement that the rebuilding was as makeshift as the original structure. Finally, an arrangement for chaining and padlocking the prisoners to rings in the floor beams was adopted. That helped some but it did not discourage the activities of the most notorious figures.

Spanish Blacky, for example, ran one of the most notorious bootleg joints in the town, and ran it at the point of a stiletto. He was a handsome, swaggering, halfbreed Mexican with a reputation of being as willing to stick a knife into a protesting customer as to get him drunk and strip him of his bankroll. One of his favorite forms of entertainment and advertising was to stand in front of his dive and throw his knife into a mark on a telephone pole across the street. That target practice helped materially in maintaining his domination over his clientele. Incidentally he was a crooked gambler and automobile thief. One stolen car finally caused his downfall.

After driving the car around the neighborhood for several weeks he wrecked it one night while drunk. After he had sobered up and found that the car was so badly damaged that it was hardly worth repair he gave it to a garage man with whom he had been friendly. When the local police noticed the wreck under repair and made some inquiry they discovered that the car had been reported as stolen. They accused the garage owner of receiving stolen property.

Spanish Blacky unexpectedly arose to the occasion with the "heart-of-gold" stuff. To the amazement of the Bowleggers, he confessed the theft and submitted to a three-year term in prison to save his friend.

Big Nell, the bootleg queen, never appeared even momentarily so out of character. Whatever else she might have been, Big Nell was consistent. She was a buxom, flashy woman who operated a beer joint, and took plenty of cash but no dirty work from her customers. If there was dirty work to be done, Nell could do it herself, and frequently did. She could shoot a gun, roll a drunk, or use her fists with the best of them. She got away with a lot of such activities until an Okmulgee phonograph sales-man was found robbed and murdered in a field near Semi-nole. Investigation revealed that he had last been seen alive in Big Nell's joint. Nell and Bob Hendricks, a gun-man hailing from Colorado, were accused of participa-tion in the crime, and convicted. Hendricks was sent up for a ninety-nine-year term. Nell was released after serving eighteen months. She was last heard of in the Texas oil fields.

Wingy McDaniel, another of the Bowlegs bad actors, kept a rooming house in which drunk-rolling and subse-quent activities with knives and guns were frequent. On one occasion a gank of drunken hoodlums engaged in a brawl in Wingy's place in which one of the participants was slashed to ribbons. With his boots spouting blood at every step he toured the night spots to show how badly he had been knifed and how he could take it. When finally he tried to get into a drugstore for treatment the druggist protested. "Hey, you can't come in here and get that blood all over the place. Stay outside until the am-bulance comes." The victim, still boasting of his tough-ness, stood in the street until he dropped dead. They were tough hombres in Bowlegs.

In one of the town's first Fourth of July celebrations rival gangs decided to use ball cartridges instead of blanks in their guns. Bullets marked store fronts and house fronts and a few of the battling celebrants while honest folk remained within doors for hours until police reserves from Seminole and Wewoka managed to quell the dis-turbance.

That sort of life quickly gave fame and flavor to Bow-legs, but the day was too modern for it to go on. With the first reduction in the flood of humanity following upon reduction and stabilization of the flood of oil the town settled into respectability. It is now one of numerous unimpressive oil towns in the Greater Seminole field.

Despite the unprecedented riches of the Greater Seminole field, the Indians upon whose original reservation the most productive wells were opened profited comparatively little. Of the two thousand Seminoles, including mixed-bloods, in the area, few had retained an interest in their lands. One of those was a woman named Katie Fixico. The first well brought in by Bob Garland was called the Fixico No. 1. That well was soon flowing five thousand barrels a day. Katie profited handsomely.

But most of the Seminoles did not profit at all. Land allotments to them had been completed twenty years before there was proof of oil in that region. Government supervision of the Indians had slackened as they began to care for themselves more or less successfully by farming. The final registration of the Seminoles had been completed in 1906. Those who had less than half Seminole blood were entirely free of Federal control. Many of them had either sold or lost their allotted or inherited lands before the oil discovery. Others were only partly restricted or protected by the government. The result, in general, was that the white men got the money when the barren lands turned into flowing reservoirs of wealth.

The career of F. J. Searight, for whom the Searight Pool of the Seminole field was named, furnishes a typical example of the methods of promotion and the occasionally fantastic success of that region. Searight, a real-estate man of Minneapolis, came into the oil country in January, 1918, and located for the moment at Beggs. In six months he owned a producing oil well. Ups and downs followed rapidly.

With what he had taken out of Beggs, Searight moved to Burkburnett. There he was impressed almost as much by the fact that men were paying twenty-five cents to be carried across a main street more than knee deep in mud as he was by the reports and enticements of the curbstone brokers and peddlers of oil leases and royalties. The fact that a twenty-five cent fee could be collected for transportation from one plank sidewalk to another indicated that money was circulating freely.

Searight decided to get some of it. He did, but lost it, and moved to Texas and back to Bartlesville on continued ventures. Come hard, but quickly, and go easy, was the usual course of the independent oil promoter's money.

When several dry holes had been drilled in the Seminole country to the discouragement of the wildcatters and their financial backers, G. E. Rogers, then of Okmulgee, came into Searight's office one day with a lease on 1,050 acres of Seminole county land at a bonus price of only six dollars an acre. Searight was familiar with the terrain, and had convinced himself, as a wildcatter or promoter must, that there was oil there. To give himself a technical opinion which might help him finance the venture he sent a geologist to check the land specified in the lease. On the strength of that opinion he closed the deal and started a test well.

It proved to be the most expensive drilling operation which the erstwhile realtor had undertaken. Lack of funds stopped work several times. Friends warned the promoter that he was throwing good money after bad, and that their cash would not be included. The scattered dry holes through the country had proved it to be worthless, they said. Drilling was extremely difficult and costly. The sands and clays and rocks all indicated that the geological formation was wrong for oil. But they did not indicate it to Searight. He persisted. And the first fifteen-day production from his discovery well alone brought him a check for $89,980. The Searight Pool was not only discovered but proved, and added its stimulus to the whole vast Seminole rush.

Incidents of that kind stimulated oil promotion. All a would-be oil magnate needed to do in the days of the first wild excitement of Seminole was to obtain a lease on a block of land in or near the area, have a blueprint made with a derrick drawn in at the site of his proposed well and set the lure for business on the sidewalk. A blueprint tacked on a board seemed to exert a strange fascination upon passersby. It carried an impression of authenticity backed by engineering and geological experience.

To the suckers especially, of whom thousands crowded into the booming oil centers, it was convincing evidence that there would be a well on the spot indicated, and if there was a well there must be oil. Many an independent well was financed in precisely that way. Royalty rights in the proposed well frequently could be sold for enough to finance and drill the hole. Waiters in the overcrowded restaurants, girls from the gaudy dance halls, truck drivers stuck in the mud within sight of one of those alluring blueprints, gamblers, traveling salesmen with a line of soap, drygoods or notions for the false-fronted stores, stopped, listened, put up their money. For the moment everyone was rich in imagination. A few became rich in fact.

Even accident and threatened disaster, if it were sufficiently spectacular, helped to spread publicity and attract more independent small-time speculators into the field. Of such publicity the most self-evident to every stranger within a radius of miles, literally a pillar of cloud by day and of fire by night, leading on with the proof of riches, were the great fires which broke out in the district. A blaze which took a toll of $2,000,000 from the company that owned the property helped to convince other speculatively inclined persons that the oil, the equipment and the profits were there to be enjoyed if not burned.

Such fires as that which raged for five consecutive days at the Hearn No. 1 of the Twin State Oil Company, seven miles north of Seminole, in November, 1927, were news widely circulated. The battle against that fire was of epic proportions. It was as full of human interest as it was of oil. Hundreds of men, tractors and teams were rushed into the battle to construct huge earthen embankments around the burning material. The blaze was being fed at the rate of two hundred barrels of gushing crude oil and fifteen million cubic feet of gas per hour. One may rest assured that the quantity of production of fuel for those flames from a single well was not overlooked in the news stories wired from the scene.

As the scrapers raised embankments as close as possible to the roaring flames amid the thick billows of black smoke, gangs of men moved up to complete the barriers. The men operated three in a group. Two held a sheet-iron shield against the blasting heat while a third wielded a shovel to move the earth closer and higher against the spread of burning oil. Steam from twelve great boilers was poured into the inferno. The flames lighted the countryside by night, and the smoke obscured the sun by day.

For five days that warfare went on, with the fire and the story of the fire growing greater and gaining wider publicity each day. Nationally known oil well fire fighters were brought from half a dozen states to direct the battle. It was a thrilling story.

And when M. M. Kinley walked into the inferno behind an asbestos shield, carrying thirty quarts of nitroglycerine, and blew the fire out after five days of Vesuvian display, the news went round the oil world. It was featured in metropolitan newspapers from New York to Los Angeles. If the Seminole field needed any wider publicity than it had previously enjoyed, the Hearn well provided it.

CHAPTER XIX

Osage Wealth, Folly and Death

ONLY in name was the drama of oil in the great Seminole field a drama of the Indians who had been reduced to that comparatively small reservation half a century earlier. In that the situation contrasts sharply with the experience of the Osage tribe where the riches of oil have been a continuous influence for a third of a century.

As explained in an earlier chapter, the Osage tribe continued to hold its land in common after individual allotments of acreage had been made to other tribesmen. Leases of Osage land for drilling purposes could be made only with the approval of the Interior Department. Any profits from such leases must be paid into a common fund for equal distribution among originally assigned "headrights" of all members of the tribe, of whatever age or condition.

The only inequalities of wealth among the Osages came about slowly through death and inheritance. In the very early days of oil development in the Osage reservation the individual incomes were small, as oil production itself was small. Later, when proof of the existence of oil in commercial quantities gave added values to Osage lands, the Interior Department required prospective leasers to submit sealed bids specifying the cash bonuses they would pay for royalty leases on various tracts. Those bonuses generally were small, and the Osages did not profit greatly, although their incomes improved. An open competitive auction system of disposing of leases was finally instituted by arrangement between the Interior Department and the Osage Council. The first public auction of Osage leases was held at Pawhuska, original capital of the Osage tribe, on November 11, 1912.

Colonel E. Walters, official auctioneer for the tribe, and incidentally to become the most famous auctioneer of oil leases in the world, with a record of having sold more than ten million dollars' worth of oil properties in a day, disposed of 107,000 acres at the first auction for a bonus of $39,000. That was approximately thirty-seven cents an acre. Ten years later he sold a third as much acreage for bonuses aggregating $10,000,000, or nearly $300 an acre.

That increase in bonuses alone for the Osage tribe in ten years is some measure of the increase in the individual income and wealth of the tribesmen. It was only a fraction of their income. The one-eighth royalties on an oil production which made experienced companies willing to pay millions for the privilege of drilling and producing was far greater. In the years of their greatest riches, in the early '20's, each and every Osage on the tribal roll of "headrights" was drawing an income of more than $10,-000 annually. Some, through inheritance, were collecting more than $100,000. Their income had been increasing through a score of years.

The few slight downs which occasionally had interrupted the higher and more frequent ups of income had made no permanent impression upon them. Under the government's restrictions they could not dissipate their capital. What matter if a buck or squaw with an income of $50,000 or so a year wrecked a $5,000 custom-built Cadillac in a ditch and bought another rather than wait to have the wrecked car repaired? What matter if a corrupt guardian or shyster lawyer cheated them out of scores of thousands? There was still far more coming in than most of them knew how to spend intelligently. And why save when the experience of a generation had indicated that there would always be more and more and more, without saving?

After more than a decade of rising incomes and increasingly profligate spendings, with logical if demoralizing sociological and economic effects, they were richer than ever. When the Marland Oil Company opened the famous Burbank Pool in Osage county in the spring of 1920, that popular belief in the inexhaustibility of financial good fortune was given new and more positive assurance than ever.

If evidence were needed it could be found in the action of the country's leading oil companies on any auction day when a new block of leases was to be put on sale at Pawhuska. On such a day a Pullman parlor car train of many coaches rolled into the Osage capital from the oil capital of the world, the city of Tulsa. A drama of millions of dollars was staged in a shabby little theater of Pawhuska.

Colonel Walters mounted the stage with maps and schedules of properties as his props, and a clerk at hand. The dingy rows of seats were filled with a motley throng. An Osage brave wrapped in a gaudy blanket might be sitting beside a Boston dowager in a mink coat, pausing for an hour on a transcontinental tour to enjoy a thrill in the scene of a million dollars changing hands at a nod. A gray-haired little man representing wealth incalculable, perhaps the wealth of Standard Oil itself, might occupy a seat beside a squaw in a blanket or an opera cloak imported from the Rue de la Paix.

"Tract sixty-five, on the east side," calls the auctioneer. "No bids less than five hundred dollars."

"Uh!" The Indian in the gaudy blanket grunts under his breath. The dowager draws her mink coat closer. The little gray man looks bored. East-side tracts are not interesting. The great oil pool of Osage county at the moment is in the west, at Burbank.

"Five hundred," says a mild voice at the rear.

"I'm bid five hundred. Who'll make it six?" calls the auctioneer. "Five, going to six! Fi'-t'-six; fi'-t'-six; who'll

Outdoor auction sale of oil leases of the Osage tribe at Pawhuska, Okla., after the business outgrew the little theater which was originally its scene. The auctioneer in the foreground, wearing one of the brilliantly striped silk shirts popular twenty years ago, is the famous Colonel E. Walters. The tree at the right came to be known as "the million dollar tree," because of huge deals, often running into more than one million dollars for a single lease, completed under its shade. The building in the background is the Indian agency.

(Courtesy *The Oil and Gas Journal*)

make it six? Thank you," he nods to another man who
has lifted a hand. "Six hundred dollars; who'll make it
seven? Six-t'-s'en; six-t'-s'en; give me seven. Gentlemen,
we have a lot to do here today. Going at six; once; going
twice; three times! Sold, at six."

There is nothing very exciting about that. It is in fact
rather dull. But as the day wears on the drama improves.
Bigger and more important oil men have entered the theater as richer tracts come up for sale. The crowd begins
to move to the edge of the seats. A sigh of abruptly broken
tension sounds as Tract No. 13 is knocked down to the
Sinclair Oil and Gas Company for $615,000.

"Tract fourteen," the auctioneer announces. "What
am I offered for tract fourteen?"

It is not necessary to tell these men anything about
tract fourteen, or any of the others now coming to the
block. They know the geology. They know the production of wells in adjacent acreage. They have estimated the
cost of every detail of development, production and marketing. There is almost no limit to what they could pay
for that lease, but there is a very definite limit, based upon
their figures of costs and potential profits, as to what they
will pay. If they can buy the lease more cheaply their
profits will be higher; that is all.

"What am I offered for tract fourteen?" the auctioneer
repeats, and turns a steady eye upon a single man near
the center of the floor. He knows that man and the money
he represents.

"Five hundred thousand dollars," comes the answer.
The spectators edge forward.

"Five hundred thousand; who'll make it six?" The
auctioneer points a straight finger at another man in the
group. He knows them all and the accuracy of his knowledge brings prompt results. A nod answers the pointing
finger. "Six hundred thousand dollars for tract fourteen,"
he says. "Am I offered seven?" He turns to the first bid-

der and notes a nod. The small fry have dropped out. This is a battle of giants. They are calm, gray-haired, steady-eyed giants of the battle for oil which has been going on in the United States for more than half a century. They are the least excited persons in the theater.

"Seven," says the auctioneer. "Do I hear eight?"

"Eight," says the second bidder.

"Nine?" Colonel Walters looks and notes another nod.

"Nine it is; nine hundred thousand dollars for tract fourteen. Do I hear a million?"

Another nod from the gray man who does not even lift his eyes from the map spread upon his knees.

"A million dollars is bid," says the auctioneer, and a nervous ripple of applause sounds from spectators who get a thrill even at the thought of that much money.

But the limit is approaching. "One million; who makes it eleven hundred thousand?" There is no answer. A sigh of disappointment goes through the crowd. "And ten thousand," says one of the contestants. Slowly, by additions of ten thousand each, the price works up to eleven hundred thousand dollars. "Eleven, now twelve," chants the auctioneer. "You know this is Burbank, gentlemen. Don't overlook your hands." There is only a suppressed sigh of disappointment from the audience. Colonel Walters knows the signs. No need to waste time. "Going at eleven hundred thousand," he says. Once, twice, and sold!

The whole room relaxes with a sigh. A few minor leases are knocked down at lesser prices. Then a renewed tensity appears. Another tract starting at half a million mounts to $1,101,000 and goes to the Skelly-Phillips interests. Everyone except the bidders is on the edge of the seats again. Tract 21 starts at half a million and mounts quickly to one million dollars. Advances drop to $25,000 each, but come in rapid fire until the total is at $1,200,000. The bidding stops. A hush reveals the tension, waiting for the sharp word, "Sold!" But instead a low whistle from near the rear of the room turns every head. A hand is seen extended, with five fingers spread.

"And five thousand more," says the auctioneer. There is a new thrill, more perceptible than a sound, through the room. The race, bogged down in the last furlong, starts again. And five, and five, and five, the auctioneer acknowledges.

"I have twelve hundred and fifty thousand offered," There is no answer, no word, no signal. The crowd awaits the hammer. Colonel Walters makes a final effort, pointing to the original bidder. A nod brings a grin of satisfaction. Walters has played with the Osage boys in days of poverty. He will get them all he can. "Twelve hundred and fifty-five thousand dollars offered for tract twenty-one. Who adds five thousand?" There is another nod and the excitement strengthens once more. Five, and five, and five, it goes.

One million three hundred and five thousand dollars is bid. Again there is a moment's hesitation. "And another five?" questions the auctioneer. Silence. But another hand appears. "It is one million three hundred and ten thousand dollars," the voice records. "And five?" he questions again. No answer. No signal. "Sold!"

The play is finished for that day. Some ten million dollars have been added to the coffers of the Osage Nation to be distributed among the beneficiaries and turned into motor cars upholstered in crimson plush and mansions which may be left unoccupied while their owners make themselves more comfortable in a teepee, or dissipated in New York, Paris or Hollywood. What of it? That is only bonus money. The royalties will continue to pour in with the development of those leases.

It is an authentically engraved invitation to folly—and to crime. At first the folly was that of children turned loose with a free hand in candy and toy shops. The crime was the crime of larceny by bailee, embezzlement and similar fraud, most frequently perpetrated by guardians, shyster lawyers, or white men and women marrying into the tribe. It moved on logically to murder. And murder became wholesale.

The body of Anna Brown, a wealthy Osage woman, was found in the spring of 1922 shot through the head, in a wooded ravine east of Fairfax. The authorities showed little interest. Indians, especially when drunk, as numerous Osages frequently were, were likely to be free with firearms. She might have been killed by accident or in a brawl. Nothing was done about it.

A few weeks later Henry Roan, a cousin of Anna Brown, was found dead in his automobile with a bullet through his brain. A cursory investigation revealed that Henry Roan had inherited a fortune from Anna Brown, that Anna Brown had inherited from Lizzie Q. Brown, an aged Osage woman who had also died without benefit of medical attention. Some weeks after the death of Roan, one Charles Whitehorn, related by blood to those previously named, was found dead near Pawhuska. Two months later George Bigheart, son of a hereditary chief of the Osages, was stricken acutely ill and removed to a hospital in Oklahoma City. With him went William K. Hale, a well-to-do rancher, and Hale's nephew, Ernest Burkhart, brother-in-law of Anna Brown.

Charles Vaughn, Bigheart's attorney, interviewed his client in the hospital. What he learned was never recorded. The next day young Bigheart died. The day after that Vaughn's body was found beside a railroad track between Oklahoma City and Pawhuska. Terror struck the tribe. It was noted that death was striking the wealthiest persons on the rolls. It seemed to come most quickly to those whose wealth had recently been increased by inheritance.

Numerous Indians fled the state. Others filled their homes with guns and ammunition, and strung electric lights around the places as an additional safeguard by night. Others were killed.

In May of 1923 a dynamite explosion destroyed the home of W. E. Smith, whose wife had inherited a large part of the Brown-Roan income. Smith, his wife and a servant girl were killed by the blast. Demands for investigation, prosecution and the protection of the Indians became so insistent that action was started.

A point of interest quickly discovered was that several combined incomes had come down by inheritance to Mollie Brown, wife of Ernest Burkhart, nephew of W. K. Hale. Hale was a power in Osage County. He had money, and spent it freely. He had even collected a $25,000 insurance policy made out to him on the life of the murdered Henry Roan. He had explained that to the apparent satisfaction of the authorities by saying that it had been part of the security arranged by Roan for a loan before Roan had inherited the Brown income. None of the local authorities seemed to want to arrest Hale or Burkhart, or otherwise incur their displeasure.

Investigation of the triple murder in the Smith home, however, did bring under questioning one Bert Lawson, who had been questioned with reference to the earlier murders. Lawson had a marvelous alibi. He had been in jail at the time of each of the murders under investigation. It seemed that nothing could be done in Osage County. The terror was complete and paralyzing. But Lawson was removed to Oklahoma City by state authorities and subjected to an examination which brought about an amazing confession.

He admitted not only the Smith murders but numerous others. He explained that he had been released from jail for that purpose in each case. All preliminary arrangements had been made by others. He was freed for the time necessary to do the actual killing, and then returned promptly to his cell. He implicated Hale, Burkhart and M. A. Boyd, an erstwhile deputy sheriff of the county.

Still no effective action was taken against the men named by Lawson. The confessed murderer was sent to the state prison at McAlester and thence to the Federal prison at Leavenworth when rumors circulated that an effort would be made to murder him or release him. Investigation slowed down. But the extent of the conspiracy and the number of murders, which had reached a total of seventeen, was brought to the attention of the Department of Justice in Washington on the plea that some of the murder victims were wards of the government.

Federal investigators and prosecutors did not have to live in Osage county. They had no terror of any organized band of murderers for hire or profit. Their greatest difficulty was in finding witnesses who would jeopardize their futures by incurring the powerful enmity of the accused men. But they obtained enough evidence to indict Hale, Burkhart and one John Ramsey, another well-to-do farmer.

Burkhart was first brought to trial in Pawhuska in January of 1926. His wife, Mollie Brown, whose income, augmented by inheritance from some of the murdered Indians, had grown to more than $100,000 annually, appeared, blanket-wrapped, in a gaudy limousine with chauffeur before the Pawhuska courthouse each day of the trial. She showed no emotion, and little interest in the trial. Reporters doubted that she understood its details or significance.

In the course of the routine of identification when Burkhart was questioned on the stand as to his business or line of work, he answered contemptuously, "I don't work. I married an Osage."

Two weeks of evidence piled upon evidence convinced the defendant that he could not hope to beat the case with all his wife's money and the Hale power of terrorism available to him. Abruptly he changed his plea to guilty and threw himself upon the mercy of the court. He saved his neck for a term of life imprisonment. He also named his uncle, W. K. Hale, as the man behind the whole conspiracy of wholesale murder. One Asa Kirby, he said, was the man who had actually dynamited the Smith home and family. One Henry Grammer was named as a go-between. Both Kirby and Grammer, incidentally, had been killed between the time of the crime and the time when they might have been called as witnesses.

But there was sufficient evidence without their testimony. There was also a persistent power of the terrorism which seventeen murders had impressed upon the tribe. When Hale and Ramsey were placed on trial in Guthrie, with the Federal government prosecuting, Ramsey repudiated a prior confession that he had shot Roan at Hale's orders. Hale himself, hard, cold, poker-faced, denied the whole story. Witnesses faltered before his steady gaze. After fifty hours of deliberation the jury was discharged.

But this was a Federal case. Local witnesses or local veniremen within the zone of terror might be coerced or intimidated but the Department of Justice in Washington was not. Promptly it transferred the case from Guthrie to Oklahoma City. And there it obtained a conviction of both Hale and Ramsey. Both were sentenced to life imprisonment. The reign of terror which had been brought upon the Osage Nation with its unprecedented riches was ended.

Civilization once more had triumphed upon the last frontier. The Osages were free to go about the business of living and dying in their own way—some by crashing their high-powered cars into other high-powered cars or into ditches and trees, some by overindulgence in the fire-water which their wealth made easily available even in Federal prohibition days, and some quietly and soberly, wrapped in their blankets.

Methods more subtle than murder were developed to divert a part of the flowing wealth from the Osage tribe to the hands of white men who considered themselves able to use it in a manner more suitable to the civilization which they represented. Some succeeded. Some failed. All were comparatively painless. There was so much money left that the Indians hardly missed the diverted fractions. Few of them even knew or suspected what was going on.

There are still technical and legal doubts about some of the practices employed. As late as November 28, 1936, the Federal government revealed one opinion, based upon investigation, by filing suit against a group of oil and pipe-line companies for sums aggregating several million dollars allegedly chiseled from the tribe. According to the charges, it was a simple bookkeeping job by which three per cent of the Indians' oil was deducted from the total "on account of dirt and sediment."

Clarence Bailey, U. S. District Attorney, filed the suits in Federal court in Tulsa. He named the Carter Oil Company, Sinclair-Prairie and Stanolind Crude Oil companies as purchasers of the oil, and the Oklahoma Pipe Line, Stanolind Pipe Line and Sinclair-Prairie Pipe Line companies as transporters. The actions were in two groups, one alleging that the purchasing companies had deducted three per cent of the purchase price; the other that the transporting companies had not reported the full amount of oil received as prescribed by law. An accounting of the oil transactions going back for more than thirty years was asked.

The Sinclair-Prairie company alone was sued for $1,-139,531.56. Whether the Osages will ever recover any of the total of millions involved in the suits remains to be seen. If they do, probably they will need it by that time. The Burbank Pool, which poured the greatest riches into their blankets, is not all it used to be. Many of the smaller pools within their original reservation are exhausted or almost exhausted. Others are declining. The Osages still enjoy incomes which would seem riches to the average American but they no longer abandon a bogged automobile because it is easier to buy a new one.

Only one, Ho-tok-moie, more widely publicized as John Stink, has persisted, without personal effort or press agent, in breaking into print through the years of decline of Osage wealth. John Stink's claim to notoriety has been based more upon the conspiracy of circumstances which have emphasized and displayed his eccentricities than upon wealth and extravagance. Quite recently an authoritative magazine of international scope devoted a thousand words to conflicting reports concerning the old Indian. It concluded with an editorial comment that John Stink's story comes in many forms. That is true. I picked up half a dozen versions from half a dozen sources in Tulsa and Pawhuska, none of which are identical with each other or with the variorum mentioned above.

The precise truth is difficult to discover, partly because John Stink though still alive at this writing has become almost a legendary character of the Osage Nation, partly because he speaks practically no English and very little Osage, and partly because he has had substantial reasons to doubt the good faith of mankind. He has been an outcast from his tribe and he is in truth a hermit. Also he is an "incompetent," under the care of a guardian appointed by the government. He has no designs on Hollywood. Probably Hollywood could not use him in person even if it could entice him, although it might contort his story into a scenario.

Anyway, the precise details make little difference. John Stink and his story are interesting because they present a picture of Osage background, and what could happen in an Indian tribe made rich beyond its average intelligence.

Ho-tok-moie was born in Kansas, probably about 1863,

when the Osages were still living there. He was a dull and backward boy and man, of no importance to the tribe, to himself, or anyone else for nearly half a century. Even when he came into Osage "headrights" with an income of a thousand dollars a month or more, he was not of much importance. He did not know how to spend that much money, and he did not care to learn. An official guardian doled out what he needed and the remainder accumulated to his credit in the tribal fund.

Most of the dole went to feed his dogs, of which he owned a loyal and beloved assortment of all breeds and mixtures. From ten to a score were always at his heels. What was good enough for his dogs was good enough for him. He slept with them, ate with them, lived with them. And then he "died."

There are disputes, still recurring, as to the precise manner of that death. Some who profess to know, and whose reports have been recorded in various newspaper articles, assert that his "death" was due to smallpox in an epidemic which swept the Osage. One youthful friend and associate, Pah-Se-To-Pah, "Boy Hunter of the Osages," a deaf-mute whose handicap placed him about on a par with John Stink as a conversationalist, and who for a time was his only close associate, has recorded that John Stink did not die of smallpox but was frozen stiff in a snowdrift. Still others maintain that the Indian was accorded his cognomen because of an especially offensive form of skin disease from which he suffered.

In any event, Ho-tok-moie was officially laid away by the Osages with the proper ritual of the tribe. And the next thing the Osages knew the body of John Stink, complete with dogs and odors, was walking the streets of Pawhuska. Evidently the Osages were not a morbidly superstitious people. They knew the man was dead. They had performed the funeral rites. His reappearance did not disturb them. They simply ignored him. Willy-nilly, John

Stink became a hermit. Occasionally he would drop into town, curl up on a sunny sidewalk surrounded by his dogs, and sleep. That course of life continued while other Osages rolled by in their limousines until one day a town marshal shot one of John's beloved dogs. Thereupon the old Indian turned the tables on Pawhuska and his fellow tribesmen by boycotting the town and its residents.

They wouldn't have anything to do with him? Very well; he wouldn't have anything to do with them. He and his dogs retired to the woods. The guardian saw to it that there was enough food and blankets provided. John Stink dropped out of the Pawhuska public eye.

One report has it that employes of the Indian Bureau built him a private reservation surrounded by a high woven-wire fence, with a comfortable house and an Indian couple to look after him, virtually as a prisoner. One says that when he declined to sleep in the house, a huge teepee with hardwood floor and polished hardwood poles was erected and occupied. In any event he was becoming a figure of legendary proportions. The dogs' grocery bill was popularly reported at $100 a week. I have been informed, though I have not been able to confirm it by anyone who witnessed the ceremonies, that the tribe eventually decided that John Stink was in fact not dead but in retirement. That story says the necessary ritualistic formulas were celebrated to reinstate Ho-tok-moie to life and tribal membership. At any rate, John Stink began to reappear from time to time, although never as a popular hero such as a man resurrected from the dead might reasonably expect to be.

He still had no more use for the Osages in general than they had had for him. His best friends were Mr. and Mrs. Whirlwind Soldier, a Sioux couple who had been kind to him. To them is given credit for teaching the old man some of the advantages of civilization, including the fact that a log cabin is warmer than a brush hut in an Oklahoma blizzard. The Whirlwind Soldiers are also given credit for introducing the old man to the delights of Christmas. He likes "The Big Holy Day." He likes especially the decorated Christmas tree and the Santa Claus feature of the celebration. He even invites boys and girls of the neighborhood to join him and his dogs at a Christmas feast, at which he distributes presents to the youngsters. It is reported that he takes a shower bath prior to that feast. Civilization is slowly destroying an authentic hermit. If John Stink could live another generation or two probably he would be riding around in a crimson limousine.

But he is well past seventy years. He has learned something in the last ten years but his time is growing short. He still has the money, or at least the money exists, credited to his headright, under guardianship. Being an "incompetent" he cannot spend it, even if he could think up ways to do so. He has no living kin, so it is hardly worth while to murder him. When he dies the accumulated fortune will merely revert to the tribal fund.

Probably most of the Osages will need it by that time. The days when oil privileges on a single quarter-section of Osage land brought a record bonus price of $1,990,000, the days when the auctions outgrew the little theater in Pawhuska and spread to an outdoor gathering suggestive of an Iowa picnic in Long Beach, California, seem to be definitely past.

The curtain began to come down slowly upon that scene of the drama nearly a decade ago. But while the Osage drama was still rising toward its heights, another, smaller Indian tribe to the westward had been brought into the oil picture. The Ponca Indians and the famous 101 Ranch were ushered into the pageant together under the direction of E. W. Marland, who was to attain fame and fortune from his part in the performance, and eventually rise to the office of governor of Oklahoma.

CHAPTER XX

THE PICTURE of Indians made rich and foolish by the rush of Oklahoma oil to their pocketbooks and heads carries with it some injustice. There were many who used their wealth in intelligent and exemplary fashion. There were others more primitive who lacked the ability to make good use of the riches but who clearly recognized and sadly deplored the menace to their lands and their traditional institutions and manner of life.

Ellsworth Collings, in his history of the famous 101 Ranch, published by the University of Oklahoma Press, quotes Running-After-Arrow, an aged Ponca Indian, observing the bringing in of the first gas well on the historic range: "Uh-h, no good, no good. Beautiful country all die now. Cattle die. Ponies die. No good, no good. Beautiful country soon all gone."

There was truth as well as poetry in that expression of an Indian's vision of the forthcoming destruction of all that he believed to be of value. The gas well which prompted the statement was the second well drilled by E. W. Marland on the 101 Ranch lands formerly held by the Ponca Indians. The first well, drilled near the ranch headquarters, had been abandoned at a depth of 2,700 feet. Seven more wells produced a vast flow of gas before the ninth drilling struck oil in June, 1911, to open the Kay county field. To E. W. Marland went the credit, and an important part of the cash which followed from that opening. Marland held a half-interest in leases on 10,000 acres of 101 Ranch lands and 4,800 acres of Ponca Indian lands, given to him by George L. Miller of the ranch-owning brothers on condition that he would do the necessary drilling. When the Ponca field had been thoroughly developed it was found to be almost precisely limited to the area of those leases. Every profitable well in the field was either opened by Marland or on a sublease from Marland.

Prior to its oil boom Ponca City was an unimposing Indian town. Probably less business was transacted there than at the headquarters of the 101 Ranch, a few miles to the south. The Indians were satisfied to lease some 50,000 acres of grazing and farm lands to the Millers for $32,500 a year, and to eke out their existence with their own garden patches, hunting, fishing and occasional odd jobs. They had been induced by Colonel George W. Miller and his son, Joe, many years earlier, to accept that reservation from the Federal government, and had always been on the most friendly terms with the Millers. Because of government red tape they never sold to the Millers more than a few thousand acres, but at times the grazing leases extended over 90,000 acres.

That was the situation when a group of oil-wise friends in Pennsylvania sent young E. W. Marland to Oklahoma in December of 1908, to test their long-distance theory that the trend of oil-bearing sands and the natural direction of oil development in Oklahoma was from east to west. Marland, introduced by Colonel F. R. Kenney, a friend of George L. Miller, traveled straightway to the 101 Ranch. With the cordial support of George Miller he studied the rolling prairie of the ranch, examined the outcroppings and came to the decision that oil awaited only upon a drill. He had proved the accuracy of his geological theories in the eastern fields, and being willing to put up his own money and that of his backers without cost to the ranch, his suggestion was promptly approved. The Millers' long-standing friendship with the Indians was of great value in obtaining leases on areas not owned by the ranch.

When seven of the first eight wells brought in a tremendous flow of gas, but no oil, the 101 Ranch Oil Company, organized by Marland, with George L. Miller and a few Oklahoma men associated with the Pittsburgh capitalists, obtained a domestic gas distribution franchise in Tonkawa, fifteen miles west, built a pipe line at a cost of $500,000, and sold enough gas to finance a continued search for oil.

The ninth well, on the leased allotment of an Indian named Willie-Cries-For-War, was the winner. By the summer of 1912 Ponca City was on its way to becoming an oil metropolis second only to Tulsa itself. Proof of oil had moved another step westward. A local refinery was built, and pipe lines from the field to the railroad brought more cars to the sidings than the cattle shipments had ever required.

Marland naturally assumed leadership of the oil business in the Kay County field. While the metropolis of the field grew with less excitement but more signs of stability than some other oil-made towns, Marland organized the Marland Refining Company and took over the 101 Ranch Oil Company. The Marland company built a modern refinery in Ponca City, and expanded its activities from producing into marketing. In another decade it had extended operations into several states, and for a time Marland was one of the spectacular independent successes of the industry. That phase of his career came to an end, at least temporarily, when the nationally expanding Continental Oil Company took over the assets of the Marland company, to manufacture 475 branded petroleum products, operate 1,600 miles of pipe line, six of its own tank steamers and numerous charters, own 3,335 producing wells, and hold 1,500,000 acres of potential oil lands under lease. Big business—centering in Ponca City.

The Indian village and cattle-shipping center had established itself as an important oil metropolis, supported not only by the Ponca field of Kay County but by the rich Tonkawa and Blackwell areas. With great refineries and adequate transportation, with the huge central office building of the Continental Oil Company, with such men of personal, political and economic power as E. W. Marland, Louis Haines ("Lew") Wentz, Dan J. Moran and others active there, Ponca City played and is still playing an important part in oil's development of the last frontier. Drama of more than one kind has centered in Ponca City.

Almost as soon as Marland's persistent drilling had proved that there was oil in the region, another young man who was to make a similar success, and incidentally to become Marland's bitter enemy after some years of association, arrived on the scene. This bold youth was Louis Haines Wentz, born in Mt. Vernon, Iowa, reared in Pittsburgh, Pennsylvania, in a family of seven children.

Lew Wentz was a strapping athletic youngster with a talent for organization and leadership. A skillful and enthusiastic baseball player in the Pittsburgh high school, he carried baseball into subsequent political activities in his home ward. As manager of the local team he became a youthful politician of parts and power. By the time he was twenty-one he had brought himself to the favorable attention of influential men in his community.

Pennsylvania friends who had confidence in his ability, sound judgment and thrift induced him to handle a piece of oil business for them in the growing excitement of Oklahoma's oil production. Though the pay was to be small, the promises and possibilities were great. Lew Wentz arrived in Ponca City in January, 1911.

He attended quickly and satisfactorily to the business mission which had brought him there. That completed, he looked for opportunities of his own. He met E. W. Marland, who was then busily acquiring oil leases for the 101 Ranch Oil Company, and learned something of the possibilities and technique of that phase of the business. Whether the subsequent differences between the two men grew out of clashes over this leasing activity, as some observers report, or whether it was due to politics or merely incompatibility of character and temperament, is beside the point.

Possibly it was a little of all three. Wentz was a shrewd and thrifty person, disinclined to both extravagance and display. Some Oklahomans put it flatly and vulgarly that he was tight. Possibly he was in personal affairs, although ample evidence of his generosity in other ways was to accumulate. Certainly he always lived most modestly and unpretentiously. Marland, on the other hand, enjoyed the display of his success. Whatever the cause of their antipathy, the two men soon came into active opposition. But despite Marland's greater popularity, Wentz evidently had a way with him.

When the Blackwell field northwest of Ponca City began to excite the citizenry Wentz was able to induce numerous farmers of the region to give him leases on terms which he could finance from his slender resources. At times he saw his last dollar in jeopardy, but he managed to maintain solvency and to persist in his leasing and promotion work. When the Tonkawa and Blackwell fields reached their peak production, the Wentz share of that oil amounted to twenty thousand barrels a day.

Careful avoidance of waste and other errors, shrewd investigation, constantly improving practical knowledge of the oil business, hard-headed determination to hang on and go through with any proposition to which he had committed himself, eventually brought its reward in millions. The business, even when it was difficult as it was in the early days, did not blunt Wentz's sensibilities any more than his ultimate success inclined him to personal display or extravagance.

Rex Harlow, of Oklahoma City, author of numerous thumbnail biographies of Oklahomans, recites an illuminating incident. According to this story, Wentz noticed that many children of oil workers and the poorer residents of Ponca City with whom he came into contact were without the toys which he believed were a proper feature of childhood. The very fact that a busy oil promoter could notice such a thing is a revealing sidelight on his character. Perhaps it had come from his youthful experience in a working quarter of Pittsburgh, where he had had his lesson in the high value of recreation among the children of the poorer families. The need seemed much the same in some parts of Ponca City. Wentz had little money at the time, but he had established credit at one of the local banks. He asked a local merchant if there was a stock of toys left over from the Christmas trade.

"Too many," the merchant told him.

Wentz made a deal to buy the whole lot. Then he went to his bank and borrowed sufficient money to pay for the toys. Without permitting his name to be known in the affair he arranged for distribution of the toys among the children. One of the citizens on the committee of distribution was the lending banker, ignorant of the source of the money which had paid for the toys. As the years went on and Wentz became more prosperous, similar distributions of toys to poor children became an annual event of thrilling importance to the youngsters of the city.

Toys were not the only items of free distribution financed by Lew Wentz, according to Harlow's records. On one occasion word was carried to the superintendent of Ponca City's schools that private funds were available to provide shoes and stockings for all the needy children in the city if the superintendent would check the need and manage the distribution. The superintendent attempted vainly to discover the identity of the anonymous donor, but finally gave it up and agreed to act, even in ignorance of the source of his supplies. Never since then, says Harlow, has any needy child gone unshod in Ponca City or its environs.

As Wentz's fortune mounted the extent and variety of his philanthropies mounted with it. Always he investigated and planned each detail with meticulous care, precisely as he planned and executed his business affairs. When he conceived and promoted a hospital for crippled children, largely endowed by himself, his first care was to use his influence in the legislature to obtain a comprehensive state law under which the hospital would be intelligently and permanently administered. When special operations requiring special skill are required, Wentz can always be counted upon to provide any extra funds needed.

Husky normal children as well as the unfortunate benefit from his philanthropies. He has built and endowed a camp for the Boy Scouts of Ponca City at a cost of some $200,000. He has established a loan foundation for poor but ambitious students at the University of Oklahoma and A. & M. College, so ably organized and administered that defaults on loans are rare indeed. Ponca City oil has been turned to good use in the processes of civilization by Lew Wentz.

The third of the three figures which are being sketched in the effort to reveal the development of Ponca City's part in the oil romance of Oklahoma is of still another type.

Dan J. Moran was born to the oil business. His father was a field man for the Standard Oil Company. With the advantage of a specialized college education, which comparatively few men of the early days of Oklahoma's oil development enjoyed, he found a job in the Glenn Pool at its start. Soon he was a trusted employe of the Texas

Company. Around Bartlesville, in Texas, Mexico and Central America he extended his experience and practical knowledge of oil, of how to get it and what to do with it. Simultaneously he improved his knowledge of the human characteristics of the men who discovered it, produced it, refined it, transported and marketed it.

In his school days and college days Dan Moran supplemented theoretical and technical education by employing his summer vacations in practical work as telegrapher, gauger, pumper, connection foreman, rig helper, tool dresser and field clerk. As one of a family of nine children he learned the practical art of human relationships. When he was graduated from the Case School of Applied Science he was thoroughly equipped to apply science, labor and personality alike to his job. His advance to responsibilities of field engineering, refinery construction engineering and so forth was as swift as it was inevitable. Twenty years after his entry into the business he was president of the Marland Oil Company. With the Marland's absorption by Continental the following year he moved into similar responsibility in that vast corporation.

There is a story attached to that deal which throws an illuminating light upon the character and ability of Dan Moran. Mergers of many millions of dollars, of companies holding a million and a half acres of oil lands under lease and producing oil from more than three thousand wells, are worth a story. This one has a human element which makes it a favorite tale of Claude V. Barrow, oil editor of the *Daily Oklahoman*. As Barrow told it to me the proposed merger was held up in New York by directors of the Continental Oil Company who questioned the value of assets of the Marland company. The House of Morgan, heavily interested in Conoco, was in favor of the merger, but the Conoco directorate was not unanimous. Some of the directors demanded to be shown. Dan Moran undertook to show them. He invited the whole group out to Ponca City to make a tour of the oil fields, oil wells, refineries, pipe lines and other properties of the Marland company. It promised to be an interesting junket for the New Yorkers. They accepted the invitation.

When they descended from the train at Ponca City, the baggagemaster had more work to do than he had ever encountered with the arrival of a single train before. Trunks, gladstones, golf bags, cases of guns and fishing tackle, overnight bags and so forth made a small mountain on the station platform. Dan Moran looked them over. There must have been a twinkle behind his eyes if not in them.

"Gentlemen," he said, or words to this effect, "I am delighted to see you. I am delighted to recognize this evi-dence of your understanding and appreciation of the pos-sibilities of life in the West. I am sorry to disappoint you for the moment, and will be happy to show you the best of golf courses, trout streams and shooting country at your leisure. But first we must have our business trip. The Marland company is operated on a strictly efficient busi-ness basis. Its assets cover a large area. I assume that you have accepted our invitation primarily to inspect those assets. Kindly designate the bags which you will need for a few overnight stops and the porters will load them in these cars."

And before the jovial representatives of Wall Street on a holiday rightly realized what was going on they found themselves, with a single bag apiece, headed out through the Marland wells, refineries, pipe lines and so forth around Ponca City, onward through the other fields of the state in which it had holdings, on into the Texas fields, on into New Mexico, on into Colorado, back into Kansas and back again into Oklahoma.

Days of bumping over dusty or muddy roads, climbing out of cars at a hundred stops a day to examine producing wells, refineries, tank farms or pipe lines, nights of discomfort in outlandish hotels or bunkhouses, hasty meals at tables with oil-smeared and sweat-soaked workmen at company boarding houses followed with impressive speed and monotony. Dan Moran was unrelenting. They had questioned the extent of the Marland assets. He would show them, to the last derrick on the last acre of leases in the entire mid-continent field.

Soiled and disheveled, stiff and sore, red-eyed and hollow-eyed with weariness in the unaccustomed physical strain of the days and discomfort of the nights, the weaklings began to plead for relief.

"Oh, no, gentlemen! You've only seen a fraction of the Marland assets. Tomorrow we have to do three hundred miles through the last of the Texas fields and on into New Mexico. We have to see two hundred and eighty-seven wells producing in that area. There are five big tank farms, three refineries, two hundred and eighty-five miles of pipe line, twenty-seven drilling rigs in operation, eight hundred and seventy-two employes, nineteen company boarding houses, three company hospitals, one hundred and eleven thousand barrels of crude with an average gravity of thirty-seven coming out of the ground every day, and company trucking fleets and supply fleets of one hundred and nineteen cars in operation, with seven garages manned by thirty-nine mechanics to keep the cars operating at maximum efficiency and minimum cost. You must go on. This is quite an organization. You ain't seen nothin' yet."

At least it sounded that way to the visiting directors. They quailed. Some agreed that they were convinced. Some sneaked out with their bags and boarded the first available train back to New York. They would do anything, vote for anything, to get away from the terrific demonstration of the Marland company's wealth, diversification and extent which was being staged by the indefatigable Dan Moran.

The junket was what might be termed a howling success. By the time the remains of the party was dismissed at Ponca City to reclaim golf bags, guns, fishing tackle, riding breeches, dinner jackets, etc., and totter stiffly to a Pullman car bound back to Wall Street the deal was as good as completed. That, according to Mr. Barrow's delighted telling of the story, was Dan Moran—both a product and a producer of the oil business. That was Ponca City energy, effectively applied.

The fact that Marland had been practically frozen out of his own company, with a loss of a large part of his fortune, is another story, quite aside from the anecdote. Marland, somewhat like Cosden, had operated in the grand manner. He had expanded his original small interests to staggering proportions. He favored executives who could extend fame and good will for the Marland company on the polo field as well as in the oil field. He had built for himself and for the entertainment of his friends and business associates a most imposing mansion on the outskirts of Ponca City, within sight of the colossal statue of the Pioneer Woman, which was also his promotion, a monument primarily to the Last Frontier, but indirectly to the accomplishments of oil. When stronger interests put on pressure, Marland toppled.

Marland has not regained his wealth and power in oil, although he is still interested in the business. His comeback to influence and prominence in the affairs of Oklahoma is in the field of politics. When his governorship is finished he will be ready for more oil.

In any event, with the Continental's absorption of the Marland company another step had been taken in the organized development of Oklahoma's oil industry. Everything was ready for the next spectacular demonstration—of nature, man and money. It was a startling climax to thirty years of westward movement of oil development in the state.

CHAPTER XXI

Oklahoma City Gushes In

WHILE Bartlesville, Tulsa, Muskogee, Cushing, Seminole, Ponca City had been rising from frontier villages to cities on the westward tide of oil discoveries, Oklahoma City had been forced to make the best of its opportunities in other lines. Most of its residents knew of oil only by hearsay. They were busy building their city on a broader foundation of agriculture, shipping and general trade and industry.

To be sure, L. C. Hivick had drilled a well in the county in 1903, and had actually struck oil at a depth of 1,900 feet, but was able to produce only five barrels a day over a seven-day period by bailing. He recalls that there was an abandoned well in the county at that time. Very soon there were two. Hivick moved to the Red Fork and Glenn Pool district, far to the eastward, where eventually he figured in the organization of McFarlin's and Chapman's McMan Oil Company, previously described.

A few other tests were made in the Oklahoma City region, notably by Tom Slick before his sensational success as a wildcatter in the Cushing field, by the Merchants Oil and Gas Company, by the Mutual Oil and Gas Company and by W. R. Ramsey. With other individuals and companies taking a chance from time to time, a total of twenty-seven test wells had demonstrated to the residents of the county that they had better go on with their farms and dairies and the industries and commercial activities of the state capital.

Several of the test wells had been drilled to a depth of more than 2,000 feet. W. R. Ramsey's dry hole had been abandoned at 3,990 feet. That was expensive prospecting. Such depths could not have been attained at any cost with the crude machinery which had opened the Glenn Pool at 1,200 feet. Time, experiment, failure, success, broadening markets, improved equipment, wiser geologists were all helping oil production elsewhere in the state but not all of them together had revealed an oil field in Oklahoma county.

The prairie metropolis, founded at the signal of a gun on the first day of the first spectacular land rush, April 22, 1889, had grown to importance upon resources plainly visible upon the surface of the earth. Ready to improve and profit by those resources were representatives of the best type of hardy pioneers who had cleared most of the area of the United States for civilization.

Through the next twenty years those men, assisted by a rising generation trained in the best traditions of western America, expanded and improved a cluster of tents and shacks into a small city which was a credit to the Last Frontier. It became a business center of importance to all of central and western Oklahoma. It waged and won a bitter battle with the town of Guthrie to become the capital of the new state of Oklahoma.

It was out in front. It did not need oil. But when Robert Galbreath, one of its early business men, happened to see the first small oil well at Red Fork give the first oil stimulus to nearby Tulsa, the hand of Fate signaled to Oklahoma City.

Galbreath promptly went wild with enthusiasm for the oil business. He interested Charles F. Colcord and C. G. Jones of Oklahoma City, with whom he had been associated in various business dealings. Partly with Colcord and Jones money from Oklahoma City Galbreath brought in several wells on the John Yargee farm west of Red Fork. Charles F. Colcord came definitely and profitably into the oil business.

Anything that gave Charles Colcord a profit indirectly gave a profit to Oklahoma City. He was that sort of citizen. From that early day onward Oklahoma City was conscious of the potential wealth in Oklahoma's oil, although it appeared to be too far away from the sources of that oil to profit directly as the smaller town of Tulsa was preparing to profit. The best its more substantial citizens could do at the moment was to invest in oil development far to the eastward and use their profits, if any, for development of their home city on a wider basis of agricultural, industrial and commercial activity.

Colcord was a leader in that. He was a true pioneer of the best type in American history—sturdy, courageous, daring, industrious, honest, shrewd, loyal. Men of his character, quality and breeding have lifted the territory of the United States from savagery, developed its wealth of agriculture and minerals, built its cities, and formulated the best traditions of its government.

Through a score of years of the substantial growth of Oklahoma, from the days in which it was in fact the Last Frontier of America to the days in which modern buildings towered into the sky to look down upon the abandoned teepee sites of the Indians, Charles Francis Colcord was looked upon by hundreds of persons who knew him personally and thousands who knew him by reputation as the finest living example of the southwestern pioneer.

A suggestion of his background has been given. Because of his important part in the building of Oklahoma City, and his long association with the oil business which eventually was to break out there in the most spectacular manner of the entire state's history, a more complete picture should be given here.

Colcord's ancestors came from England to New Hampshire in the very early days of the colonial settlement of America. His grandfather had followed Boone's Trace into Kentucky and established himself upon a frontier

The late Charles F. Colcord, pioneer cattleman, oil man and city builder of Oklahoma, widely honored as the finest and most successful type of American pioneer.

plantation in Bourbon county. He, like his forebears, was
a bold, shrewd, hard-working man, equally competent as
planter and Indian fighter. Also he commanded material resources superior to many of the Kentucky pioneers.
He owned slaves, worked them intelligently, accumulated
considerable wealth for a day more than a century ago
and expanded his land holdings. On that plantation,
which his father inherited from his grandfather, Charles
Colcord was born, August 18, 1859. His mother was a
Clay, of the illustrious Kentucky family of that name.

After the War Between the States Colcord's father sold
his Kentucky holdings and moved to a sugar plantation
above New Orleans. The boy contracted malaria, and
was sent to a cattle ranch near Corpus Christi, Texas,
owned by one of his father's friends of ante-bellum days
in Kentucky. He took to the hard life of the ranch like
a Boy Scout to a wienie-roast. In two or three years he
was as tough a young hombre as any on the range.

When his father came to take him home for a continua-
tion of the schooling which had been interrupted by his
stubborn illness the boy recognized a crisis in his life. He
packed such things as experience upon the range told
him he would need, crammed them into saddlebags,
picked up a Winchester, and galloped away into the night.

He was a handy kid with horse, rope and branding-
iron. He easily found jobs on Texas ranches where he
completed his education as a practical cowboy. When his
father came again to Texas the boy enticed him into what
was then a highly important part of the cattle business—
the movement of herds over the long trails to railroad
shipping points in Kansas. When he and his father moved
from Texas into western Oklahoma and organized the
Jug Cattle Company, he went back close to the first prin-
ciples of the frontier with a home in a dugout and days
frequently running into sixteen hours in the saddle.

The Jug Cattle Company joined the Comanche Pool,

an association of cattlemen with herds grazing over hundreds of thousands of acres of Indian lands along the Cimarron River. The area involved was so vast that 150 cowboys were needed to ride the lines and do the other jobs of the association. It was in that period, when the boy was only nineteen years old, that he experienced his first taste of Indian warfare, engaging in the campaign of the cattlemen against the last uprising of the fighting Cheyennes.

All that was part of Charles Colcord's background. When the first land rush into Oklahoma came in 1899 he was fully equipped to take a successful part in it, and to establish himself in Oklahoma City as narrated in an earlier chapter. The time was to come when his sound conservatism and shrewd thinking were not only to add millions to the city's values but to save millions for other investors.

It has long been the practice of American cities growing upon areas unlimited by natural barriers to permit their older districts to degenerate while more modern structures extended the city and increased values away from the point of origin. Aware of that tendency Colcord deliberately set about to maintain values by promoting and erecting modern buildings in the section of the city which originally had been most valuable but was threatened by loss of those values as business sought improved structures elsewhere. Oklahoma City's Colcord Building, Commerce Exchange Building and Biltmore Hotel are monuments to that intelligence. Those developments of course came in the comparatively recent years of the city's remarkable growth.

In the earlier days Colcord organized a real-estate and insurance business with Galbreath and Shelly as associates, invested $50,000 of his own capital in city real estate, and went about the work of expanding and enriching his town as well as himself. When the battle started for permanent location of the state's new capital, Colcord brought all his influence to bear both in Washington and in Oklahoma. It was a strenuous fight, with the territorial capital of Guthrie making a valiant effort to retain its position, and Oklahoma City won. By that time it seemed to have everything except oil to make it the state's leading city.

In the succeeding years increasing wealth from oil flowed only indirectly into Oklahoma City to help finance its steady and substantial building on a broader foundation than that enjoyed by the typical oil towns.

Year by year the oil prospectors and wildcatters were moving westward. Year by year actual oil production was coming closer to the capital. The repeated failure of test wells in the capital area had emphasized that city's necessity of building upon other resources, but the oil millions which were being produced elsewhere in the state had made Oklahoma City eagerly oil-conscious. Its business leaders knew that the oil business had grown within a human life span from Colonel Drake's first tiny Pennsylvania oil well originally financed with $1,000, to a total estimated investment of $13,000,000,000.

Oklahoma City had been preparing, consciously and subconsciously, for a quarter of a century. And then came the climax of the drama of Oklahoma and its oil. The scene was a commonplace dairy farm, a few miles south of the capital, owned by a pleasant, industrious, commonplace couple of Bohemian peasant extraction.

Vincent Sudik's father had come from Prague in 1872 and settled in Omaha. Six brothers were born in Nebraska and reared on a Nebraska farm in the thrifty, hard-working, practical tradition of their parentage. Vincent and his wife, Mary, moved from Nebraska in 1903 and purchased 160 acres of land near enough to Oklahoma City to find a market for their dairy produce. They

worked hard, improved their stock, earned some profits
from year to year, and reinvested in more land. That was
merely logical for a man and woman of peasant extraction
and farm training. They had earned the money, the land
was cheap, and there were growing children who were expected to utilize it for their own families in due time.
They did not know, or care, anything about bonds, stocks
or oil.

They went calmly about their affairs, rearing their
children and their heifers, selling their dairy products and
incidental farm truck in nearby Oklahoma City, putting a
little money in the bank in good years to finance their
needs in bad years, and withal conducting themselves
with the dignity, propriety and self respect of substantial
farm folk. In 1926 they purchased the third addition to
their original acreage.

They were not at all excited when a stranger appeared
at the farm one day and suggested that they permit the
sinking of a test well on an oil lease. They had heard of
numerous test wells which had failed in the preceding
twenty years.

But this man offered them money. They understood
money. They knew how hard it was to get by raising
fodder and milking cows. So they sold the lease. It would
help to pay taxes, and there seemed nothing to lose. One
derrick on 160 acres could not destroy much pasturage.
The farm was doing well. They had already purchased a
cottage on S. W. 24th Street, Oklahoma City, to give their
children advantages they themselves had not enjoyed.

Then, acting on the advice of geologists and drillers
who had learned a great deal in thirty years of oil production in Oklahoma, the Indian Territory Illuminating
Oil Company under the presidency of H. V. Foster
brought a new power of wealth and machinery to make
a test to end all tests in the Oklahoma County area. The

I.T.I.O., it will be recalled, had been the first highly successful operator in the Osage reservation thirty years earlier. H. V. Foster's father had been the first man to lease hundreds of thousands of acres of Osage lands for oil production, and H. V. Foster had succeeded to his high place. He had won the wealth visualized by his father, and had consolidated and expanded his gains. The I.T.I.O. had accomplished amazing things and gained amazing riches in the Greater Seminole field, and elsewhere.

When the company's geologists explained that the only reason twenty-seven test wells sunk through a prospecting period of twenty-seven years in Oklahoma county had all proved failures was that they had not been sunk to sufficient depth, the I.T.I.O. was willing and able to believe and to prove it.

Its lease buyers acquired a block of some 10,000 acres, and purchased some lands outright. On one of these sites a well identified as No. 1 Oklahoma City was started in May, 1928. For its first 4,000 feet it met with much the same conditions revealed by its predecessors of ten and twenty years earlier. But the sand and clay and rock brought up by the drill appeared to justify the promises of the geologists who had insisted that the oil was below, if the drillers would sink far enough.

The I.T.I.O. had the money, the equipment, the skill, persistence and manpower to do it. At 4,012 feet they penetrated gas sand. No oil. They drilled on. At 4,816 feet they tapped more gas sand. Still no oil, but, more important, no discouragement.

Ruthless and destructive as great corporations and vast sums of money concentrated in a few hands have proved to be at times in the history of the world, there can be no doubt that they have supplied a power which has released incalculable ultimate benefits to mankind. No Colonel Drake, no Cam Bloom drilling for Cudahy, no Dr. Clinton or Robert Galbreath could have accomplished what the accumulated resources of science and wealth of the I.T.I.O. was proceeding irresistibly to do.

Down, down, down went the drill. And at a depth of 6,402 feet, with a roar that took the breath from the surrounding crew and the top from the derrick itself, the well shot oil a hundred feet into the air. In ten days it was producing 6,564 barrels of oil each twenty-four hours. The Oklahoma City field had come in. Even the Sudiks must have been a little excited about that, although they do not admit it. They had given an oil lease on their land, but were continuing to farm it and make their living from it.

But after the I.T.I.O., through its No. 1 Oklahoma City well, had proved the accuracy of its geologists' pronouncement, a well was spudded in on the Sudik farm. It was named No. 1 Mary Sudik. Mary Sudik, pleasant, plump, hard-working, competent housewife, was not impressed by that honor. She continued quietly about her family affairs.

Then, on the morning of March 26, 1930, the calm of that countryside was shattered by a roar that could be heard for miles. Close upon the sound came a rush of gas-driven oil that could literally be seen for miles, and soon could actually be felt for miles. There may have been wells since then in some part of the globe to compare to the Mary Sudik but there has never been and may never be an oil well of such spectacular display and prompt and international fame.

Fifteen thousand barrels of oil a day spouted from the Mary Sudik. It swamped the fertile farmlands in a flood of oil. When the prairie winds caught the black plume of gas-driven spray they spread the oil for miles across the countryside. The most competent and experienced specialists in all the army of engineers and practical oil field workers in the far-flung organization of the I.T.I.O. were called to cap the well, to control the flow and check the waste and damage.

The Mary Sudik defied them. Specialists from other companies were called to the rescue. The Mary Sudik blasted their efforts aside. In a day she was "Wild Mary." As "Wild Mary" her fame began to spread as the un-fettered production of oil spread, through Oklahoma City itself and across the prairie. In another twenty-four hours every oil-producing center in the United States knew of the "Wild Mary" Sudik. In another day it began to make the front pages of newspapers a hundred and a thousand miles away. In another day it was on the cables to Rou-mania and Persia and Russia.

Still "Wild Mary" roared and spouted, gushed and sprayed. The highest skill and most improved devices which had been developed in three-quarters of a century of petroleum production throughout the world could not control "Wild Mary."

Editors of the leading daily newspapers in New York, Chicago, San Francisco and Los Angeles sent their re-porters and photographers by the fastest available air-planes to see and picture the spectacle and to interview the smiling, modest little middle-aged housewife whose fame as "Wild Mary" was suddenly on every press associ-ation wire in the world. Day after day for eleven days of constant thrills the Mary Sudik continued on the ram-page. When at last the well was capped there was hardly a literate adult human being in the United States who had not enjoyed at least a vicarious thrill with Mary Sudik.

Mary herself was the calmest person among millions. "I can't be bothered," she told reporters who crowded in to interview her. "I'm busy with my first grandchild."

She was busy with another grandchild when I managed to get around to the Sudik cottage nearly seven years later. With the tiny baby in her arms she answered my ring at the door. Her broad face broke into a smile when I stated my errand. "Ask him," she said, and shrugged a shoulder under the baby's head toward an inconspicuous figure in a shadow of the little porch. That, I learned, was Mr. Sudik, a strong, somewhat weatherbeaten figure of a man. He was as pleasantly and politely uncommunicative as Mary Sudik herself. He was more interested in his little patch of lawn than in the fame and riches which had gushed upon the family and upon Oklahoma City only a few short years earlier.

That lawn suggested an amusing contrast to Billy Roesser's monumental imported lawn acquired in Tulsa from a slightly similar flood of oil riches in an earlier day. Some unidentified variety of wild grass had killed out Sudik's planting. Now even the new grass was dying. Mr. Sudik was distressed. The thought of importing bluegrass sod as Billy Roesser did was beyond his imagination. His problem was to make this lawn grow. He was still a farmer at heart. Whatever the extent of the fortune which had come to him through the "Wild Mary" well, he had never been "new rich." He had bought this house and planted this lot on a modest street before oil had been struck on his farms. It was good enough for him and his family then. It is good enough now. But he was worried about that lawn.

"Does it need water?" I asked when he put the problem before me.

"No; it's had plenty of water."

"Maybe there is some chemical in the water that is doing the damage," I suggested vaguely, hoping eventually to get around to the subject of oil. It wasn't such a bad shot in the dark. At least it brought forth another thought. He grinned.

"It comes from my own well," he said. "Last summer was so hot and dry I didn't want to use the metered city water for my place here. I've got four lots. So I sunk a well in the back yard and put in a pump and kept everything nice and green."

That revelation of practical thrift came so naturally and unconsciously that I merely nodded, though I wondered again just how rich the Sudiks might be; or how shrewd. "Good idea," I hazarded.

"Yes," he said, and grinned again. "When the meter reader came out he found the meter dead, but the city sent a bill for twelve dollars and a half, just the same. I went to the water department and asked how come? They admitted that their report showed the meter dead, but four lots were green so they sent the bill accordingly. I won the argument," he chuckled.

"I suppose a lot of people have tried to chisel on you since the big well came in?" I suggested hopefully.

"Oh yes, some. A good many friends came around to borrow money. We lost some money and some friends." He paused. "But some still come around to borrow more."

"But you couldn't have gotten rid of all that money that way."

"Oh, no. I bought another ranch in Nebraska." Again he smiled. "I got ninety-six dollars for my share of the crop last year. The taxes were seventy dollars. The agent's fee was twenty-five dollars. I made one dollar profit. The old farms here south of town are doing about the same." He seemed rather amused than disturbed by it.

"What do you live on, if farming is like that?"

"Oh, the oil pays a little once in a while. It's about the same from year to year—big production at low price or low production at high price."

"What do you do yourself if the farms are being worked on shares?"

"Oh, we've got a couple of grandchildren around here that keep us busy. And that lawn bothers me. I've tried about everything on it. The dandelions seem to do best."

It seemed evident that I was not going to get much information out of V. Sudik about the extent of the wealth and excitement that had come from the "Wild Mary" and subsequent rich wells on his acres under lease. "I'd like to ask your wife a few questions," I said. "How did she

feel when she found she was more famous than Garbo and
Dietrich put together?" He waved toward the living
room, separated from the narrow porch only by a screen
door.

"Ask her."

Mary Sudik came smiling again to the door when I
approached and asked her a question.

"Ask him," she said.

"He wants to talk about the lawn," I protested.

"Ask him about the hard times," she said, with voice
and eyes revealing nothing but placid, smiling contentment. Whatever they had gained, whether it was wealth
as wealth is figured in the oil fields, or simple security,
shrewdly conserved, it seemed to be enough. I abandoned
the effort to find a thrill in vicarious millions or in the
reaction of this quiet woman to the world-wide notoriety
which was once thrust upon her.

Sudik was almost a match for her. He admitted merely
that there had been hard times. He declined to elaborate.
When the news of the gusher went round the world, he
admitted, he received thousands of letters, mostly begging.
Churches and organized charities pleaded for endow-
ments. Numerous couples wrote to flatter the happy
domestic success of the Sudiks, and to beg money with
which their own matrimonial hopes might be promoted
to similar success. One woman in the old country sent
five dollars to be invested. She said she needed shoes but
hoped Mr. Sudik would invest the five dollars so that it
would make her a fortune. He sent it back with advice
to buy the shoes.

With that sound suggestion we may leave the Sudiks.
Peace, contentment and security to them, to their children
and to their grandchildren. If there were more such conservative beneficiaries of the wealth of oil in this country
there might be less oil, but probably there would be more
security, without the necessity of a Federal program to
assure it.

CHAPTER XXII

Big Money Poses Hard Questions

It is a fact well known to oil men, never emphasized to small potential speculators in oil promotions, that vast sums have been sunk in the earth to produce sums hardly greater. In that the history of oil financing and profits closely parallels the history of gold and silver mining. A comparatively few individuals have profited greatly, by luck as with the Sudiks, by daring as with the late great Tom Slick, by a combination of luck and government control as with the Osage tribe or by knowledge and shrewdly applied energy as with Lew Wentz of Ponca City, Frank Buttram of Oklahoma City and others whose successes have been touched upon in these pages.

In general the greatest production of oil has always come through the investment risk of great capital. The benefits have been distributed most widely through payrolls, purchase of supplies, transportation and attendant activities. No better evidence of that could be offered than a sum totaling $69,121,289 expended by the I.T.I.O. alone in less than ten years of its activities in the Oklahoma City field. Safe, sane and conservative citizens of the Sudik type would never have laid out that money, even if they had had it. It required men of the type and resources of H. V. Foster and his associates in the I.T.I.O.

Foster himself did not need the money. For many years he had been many times a millionaire. But there was within him an urge for expansion, for conquest of natural resources, for development and improvement and enrichment of his state. In the course of that development as revealed in the Oklahoma City field alone, the Indian

301

Territory Illuminating Oil Company in eight years distributed $11,000,000 through its own payrolls and $21,500,000 to contractors who paid the bulk of that sum to employes. It paid out $11,000,000 on leases and royalties. It expended $23,000,000 for supplies which gave profits to many manufacturers and wages to their thousands of employes. It paid $2,621,289 in direct taxes.

Altogether it distributed, outside of its own corporate profits, more than $69,000,000 which went into the hands of countless thousands of workers. Most of that was paid out through the long period of this country's worst economic depression. It maintained life, comfort and self respect for thousands without the demoralizing necessity of government relief. Yet that was only a fraction of the wealth of oil turned into cash and distributed through the operation of big business in the Oklahoma City field alone.

Other rich, powerful, competent, ambitious men and corporations came swiftly into that sensational new field of natural wealth. The first to complete a well after the I.T.I.O. discovery well roared a nation-wide announcement of the new possibilities was the Sinclair Oil and Gas Company. Nearly a month before the "Wild Mary" Sudik lifted the crowning plume upon Oklahoma City oil field publicity a Sinclair well raised a signal which was almost as effective.

It lacked the human interest element of Mary Sudik, but it was an excellent show in its own right. When the Sinclair well, blasting forth 350,000,000 feet of gas per day, caught fire, it lifted a torch such as the world perhaps had never seen outside Vesuvius or Etna. Reflection of that torch against the clouds could be seen for fifty miles. For a dozen days and nights it blazed, until at last blown out by nitroglycerine set by the skilled and daring hand of Floyd Kinley.

The late great Tom Slick also was promptly on the scene of the Oklahoma City field. One of his early wells, the No. 1 Wepaco, on the edge of the business district, ran wild and sprayed much property. A score of other wells ran wild with the tremendous pressure of gas within them. The No. 1 Sigmon released its spouting flood in a windstorm which· carried an actual rain of oil for miles and spread a black "Scotch mist" as far as the university town of Norman. By that time derricks were popping up and wells going down throughout the area.

There were thrills and riches, heartbreak and labor unequalled, in the opening year of the field. The tremendous gas pressures released from a depth of 6,500 feet, and the blasting sand destroyed the best equipment available. So-called heavy duty rigs and tools brought from the Seminole field were ruined within a few hours. Heavier and stronger equipment brought from California served little better. Manufacturers sent engineers and metallurgists to study the problem and supply the need.

These men, driven by necessity, financed by the wealth which had been accumulated from innumerable earlier oil fields, trained in the advancing skill and science which had accompanied increasing production of oil through three-quarters of a century, did solve most of the technical problems. As they solved them, one by one, they increased the output of oil to a point never before equaled. That, always, has been the American way. It is one of the characteristics which justifies a national pride.

When the first great gold rush roused the United States to a pitch of economic excitement it had never known—in 1849—the most pressing problem was one of transportation. When the Forty-niners reached the Mother Lode they picked nuggets from crevices along the stream beds with their knife points or washed the gold from sand and gravel by hand. Anyone could be a miner if he could get there. But when the surface gold had been collected in a few short years there was a new problem of digging to greater depths and of extracting the gold from

quartz in which it was imbedded. They solved that problem with improved tools ingeniously contrived as necessity
demanded, with timber-braced shafts and tunnels, with
mills to crush the rock and various constantly improving
devices to separate the gold from the dross.

When the great silver lode of the historic Comstock was
opened a decade later and the busy picks and shovels
quickly opened depths where the dangers of disastrous
cave-ins threatened to force abandonment of countless
millions of deeply buried treasure, new and efficient
methods of timbering were devised. When the heat and
flooding waters of the lower depths threatened again to
end production, new methods of ventilation and drainage
were developed.

Other problems were solved, as necessity arose, to extract the gold and silver wealth of Colorado, the copper
and other mineral wealth of Michigan, and Montana, and
Arizona. Civilized man had never been thwarted in his
effort to extract the mineral riches of the earth since the
legionaries of the Roman Empire had lashed their ore-
laden slaves up the ladders from the depths of the Rio
Tinto.

Americans were no more to be thwarted by the technical problems of production in the Oklahoma City oil
field than they had been by earlier problems in the Pennsylvania, California, Texas and other fields. The oil was
there. They would get it out. They did get it out. And
with it came new and even more difficult problems. These
were problems of profit, of marketing, of supply and demand, of waste and conservation. What profiteth it an oil
man to gain the whole pool and lose his own market?

There were a thousand factors influencing the situation.
All America was sinking into its deepest economic depression while Oklahoma City found itself starting a local
boom such as it had never imagined. At the moment it
was probably the only city in the United States excited

by visions of immediate riches beyond its wildest dreams. Promoters rushed into the district. Some of them came almost straight from prison cells occupied in penalty for other get-rich-quick schemes of a checkered past.

Many of them knew little or nothing about the oil business but they knew much about human psychology. They knew that in the excitement of an oil boom a great many persons completely lost their normal sense of values. The less such persons knew about the practical business of producing and marketing oil, the more amenable they were to the salesmanship of promoters. They had all heard a hundred tales of other persons who had skyrocketed from poverty to riches. They saw around them, in the tangible form of gushing oil wells, flaming gas producers and mounting skyscrapers, unquestionable evidence of wealth from oil. They knew little or nothing of the thousands of persons who had lost all their savings. They loved the story of Mary Sudik.

Laborers, mechanics, clerks and waitresses, and even merchants and manufacturers hurried to buy an interest in hastily-organized companies, in wells which might be nothing more than a blueprint based upon a lease upon some scrap of ground which had been overlooked or withheld from the experienced, soundly financed companies. Royalty interests, leases and fractional leases, and exciting promises in various other forms were snapped up by hopeful "investors" as fast as they were offered.

Some of the wells actually were drilled. Some actually produced oil. The fact that the odds against those oil-crazed buyers of miscellaneous promotion stocks were something like the odds of the Irish Sweepstakes made little difference. Only the winnings were publicized.

In this situation the established, conservative, experienced oil companies perceived a threat and a problem far more difficult than the scientific and mechanical problem of deep drilling in the face of gas-driven sand and similar difficulties. Most of the larger concerns such as I.T.I.O., Sinclair, Phillips, and their rivals had already experienced the ill effects of too much oil on a declining market in the Seminole field. The dangers were multiplied many fold, just as the investment necessary to produce oil in the Oklahoma City field was multiplied. But none could afford to permit a rival to extract the oil from his leaseholds through neighboring wells. They had solved that problem in part at Seminole by prorating production so that the field would not be exhausted at a rate which kept prices so low as to destroy profits.

They tried to solve it in the Oklahoma City field by agreement to stake wells only on twenty-acre units. But in two days that agreement blew up with a bang equal to the "Wild Mary" Sudik. Independents who had paid high prices for leases faced the necessity of drilling and producing at once in order to get their money back. Wells were started on ten-acre tracts. The field spread south across the line of Cleveland County and north to the city limits.

There was nothing to prevent the owner of a single lot from leasing it to a promoter who could sell enough prospective royalties to sink a well. The city, moved by the more influential oil men, restricted the drilling to a single well on each two and one-half acres, and temporarily checked the march of the derricks by zoning regulations. But the economic and political pressure was as great as the pressure of gas deep within the earth.

The injustice felt by a property owner who was prevented by the zoning system from leasing his town lot to a company which might drill a well and give him a fortune while a well upon an adjoining lot in a zone differently classified was stealing his oil and ruining his property for residential purposes, was intense and powerful: It was not long before that resentment of flagrant injustice brought about rezoning. The mud-hogs and derricks swept forward

Above: Oklahoma City's skyscraper skyline in 1937, photographed from the Civic Center Park. It should be compared with the picture facing p. 72 to demonstrate most effectively what oil has accomplished in Oklahoma.

Below: A night view of Oklahoma City oil wells in the Capitol area. Derricks are everywhere in what was recently the finest residential section of the city. Even the garden of the Governor's mansion has not escaped.

again, deeper and deeper into the city. Again they were checked by zoning regulations. Again the zones were broken down by votes of the electorate. Again the invading army swept forward upon a new salient until some of the finest homes in the city were literally daubed with oil from wells sunk in the midst of lawns and rose gardens.

In the same period the struggle between the larger, more experienced and more conservative companies and the independent leasers, promoters and producers was attaining even more war-like characteristics. The old-line companies realized that they were cutting their own throats by looting the vast underground reservoirs and flooding the market with a product which could never be replaced.

Very early in the campaign, before the smaller independents had effectively tapped the field, they agreed to shut down production for thirty days while they sought a more permanent and equable agreement for proration. When the wells were reopened production was curtailed under what was known as the Seminole "voluntary agreement" plan to forty per cent of capacity. Even at that rate the field produced eight million barrels in a year. Various forms of curtailment followed, with official state approval. And despite such curtailment production mounted to 47,795,000 barrels in 1931.

It appeared to be suicidal. Other plans were tried with state authority for enforcement. But the Champlin Refining Company, of Enid, representing independents whose purpose was to obtain any cash available, regardless of price or conservation, carried the issue to the Supreme Court of the United States and there upset the oil cart. Oil flowed as oil had never flowed before. Every well produced to capacity to prevent neighboring wells from stealing its oil. The price of crude skidded down to twenty-two cents a barrel, the lowest in the state's history.

Governor "Alfalfa Bill" Murray declared a military

zone covering the field, commissioned his cousin Cicero
I. Murray a lieutenant colonel with the National Guard
to enforce rules for restriction. According to operators
who found themselves pinched and some of the companies which found their reserves being looted by offset
wells, it was an illegal and unjust action.

Allowable production had been specified at 97,000
barrels a day. Oil in excess of the so-called allowable, produced or transported secretly, came to be known as "hot"
oil. There were some sensational revelations of such production even among the reputable companies. When
caught they made the excuse that other producers were
looting their reservoirs through offset wells.

Several operators openly defied the militia. Pitched
battles occurred. Tear gas came into play. Guardsmen
went about with loaded sidearms and fixed bayonets, and
occasionally used them. Wells were shut down and gates
padlocked. Pipe lines carrying "hot" oil were dug up.
Tanks were destroyed. The story of the war was carried
in metropolitan newspapers throughout the country. Esti-
mates of "hot" oil produced in two and one-half years put
the total at twenty million barrels.

It was folly and waste. It could not go on indefinitely.
Already some of the wells in the southern part of the field
were showing a marked decline in production. In April of
1933 a new proration law was passed, with a conservation
commission having wider powers and a proper police force
to see to enforcement.

In the meantime the march of the mud-hogs and derricks had extended into the capital grounds themselves.
They were held up only briefly while the state obtained
the most profitable contract possible for its leases. Marland succeeded Murray as governor. A well was sunk in
the garden of the executive mansion. And so it went—
and still goes.

The Oklahoma City field posed problems before the industry such as had never been posed before. But the oil business had long been developing men to solve problems. Among the monuments to their achievements had been erected such imposing modern buildings as the Philtower and the Exchange National Bank in Tulsa, the Ramsey Tower and the First National Bank of Oklahoma City and others too numerous to mention.

But the new problems of stabilizing production were complicated by human behavior, by economic and political pressure, outside of exact science. A thousand farmers and others in the state had leased their lands for oil development on a basis which required the lessees to pay a specified sum in cash rentals to hold the privilege of drilling, and a royalty amounting usually to one-eighth of the value of the oil produced, when, as, and if it were produced. The extent of that income and the power of its influence in forcing continued development is most effectively revealed in figures. In Oklahoma during 1935 lessees paid landowners $5,850,000 in cash rentals on 6,500,000 acres of non-producing leases. Farmers and others were collecting an average of ninety cents an acre from oil men on land which was not producing oil, and at the same time were using their land for any purpose they wished except oil production.

One farm in Noble county received oil-lease rentals from twelve different companies for twenty-three years before oil was found and royalties added to the farmer's previous cash income.

Some companies were paying as high as fifty dollars an acre to hold promising tracts for drilling. Obviously that was a burden upon the lessee which could not be carried indefinitely. The richer established concerns might manage it for longer periods than the wildcatters. The latter must drill for quick action, to relieve themselves of cash rentals on unproductive leases, to strike oil regardless of the market.

That is the explanation of what is known to the industry as "dry-hole money." Companies or individuals holding leases on certain acreage would co-operate to sink a test well. For example, if ten independent lessees were paying for equal rights on equal areas of some farmer's quarter-section, they might agree to contribute equal amounts of cash to drill a well at a point selected. The lessee of the site agreed upon might either put up more money than the others for the possible advantage of having the first producing well in the area, or balance the account in some other way satisfactory to his associates. If the test produced oil the adjacent leases immediately soared in value. If it was dry the lessees were relieved of the necessity of continuing to pay cash to hold the worthless ground.

All such factors continued to extend development of the fields and to promote production beyond a point where it could be marketed at a profit. It was essential to the industry that that situation should be corrected. Attempts at correction by agreement had been made with more or less success in the Seminole field, the California fields and elsewhere. But there was always the danger of infraction of anti-trust or interstate commerce laws, and another danger that unscrupulous members of the co-ordinating group would destroy the value of the plan by marketing "hot" oil.

There are ethical, social and economic factors involved in that situation which this writer does not presume to analyze or pass upon. On the surface it would appear that future generations of Americans should not be deprived of the natural resources of oil which they may need. The oil can never be replaced as timber, for example, can be replaced through competently administered reforestation. Once it is out of the ground it is on its way to complete and final destruction as fuel or lubricant. It cannot be reclaimed even in part as copper, for example, is reclaimed.

But, on the other hand, is consideration for a generation yet unborn as important as consideration of the rights and comforts of the generation now living? If an individual Oklahoma farmer, for example, has been struggling for years to wrest a living from a rocky quarter section of land, and suddenly discovers that there is oil beneath that land, who is to say that he must not profit regardless of any adverse effect upon a generation unborn?

Perhaps it all comes down to questions of the greater good of the greater number, of the rights of the state or society in the abstract as opposed to the rights of the individual. Perhaps many persons, especially in the modern trend, would maintain that the rights of the majority and the conservation of natural resources for future generations should be paramount. But if any such person actually owned a bit of land under which there had been proved to be a rich body of oil-bearing sands, and was abruptly deprived by law or agreement outside his influence of the profits otherwise assured, it seems probable that that individual would send up a great wail of injustice.

That then was the problem which the oil business in general and the Oklahoma City field in particular faced in its vast production. It was not a new problem, but it had never before been so emphasized. Out of the need of adjustment and the experience which had emphasized that need came men eager to attempt a solution. One of these men was Wirt Franklin, then a leading figure in the Oklahoma City field.

CHAPTER XXIII

OIL'S ACCOMPLISHMENTS

WIRT FRANKLIN, although born and reared in Missouri, had had long and intimate experience with the growth of Oklahoma and the development of its oil fields. That practical experience dated back to a job with the Dawes Commission at Muskogee in the preparation of the Choctaw and Chickasaw rolls on which allotments of lands to the individual Indians were based. Prior to that job he had studied law at Columbian University which was to become George Washington University. He continued the study while employed by the Dawes Commission, and was admitted to the bar.

Oklahoma was very much excited about oil in those days, although actual production and the total wealth derived from it were a tiny fraction of what they were to become. Young Franklin could not have overlooked that excitement and its possibilities if he would. With his law partners, S. A. Apple, Ed Galt and Ray Johnson, he acquired a vast amount of information concerning the oil business in general, and some money with which to apply his knowledge in the field.

Franklin, Apple, Galt and Johnson associated themselves with J. M. Critchlow in the drilling of what became the discovery well in the rich Healdton field. When that discovery had made their leases valuable, the four law partners organized the Crystal Oil Company and assigned their holdings to it. Development proceeded, and in 1916 they sold the company to the Sinclair interests for $2,000,000. At the same time Franklin, Apple and J. W. Harreld, later a U. S. senator, organized the Apple-

Franklin Oil Company, which was sold later at a profit.

Wirt Franklin was established as a man who knew the oil business through personal experience and had the money to prove it. He continued to operate independently and successfully. At about the time of the discovery of the Oklahoma City field he organized the Wirt Franklin Petroleum Company, capitalized at $10,000,000. It was generally recognized at the time as one of the strong independents in the business.

But being an independent at that time was something of a problem in a generally demoralized oil industry, as it had been to a lesser degree at various earlier times. They were fighting almost as bitterly among themselves as they were against what they called the discrimination of the major oil companies. It was a situation generally satisfactory to the majors, who recognized the disadvantage under which independents labored because of lack of cooperation. The majors usually could keep the independents down to a more or less harmless level by controlling the price of crude, transportation and credit.

Most of the independents were producers of crude oil, wildcatters who opened new fields and put the product on the wholesale market. When occasionally one of them found his production or potential production so large that it could not be marketed at a profit he might establish a refinery of his own and attempt competition with the majors along their own lines. That usually necessitated adoption of policies which the older larger companies had proved to be practical. Thus they ceased to be dangerous competitors, or actually sold out to the majors.

To correct that situation, to give each independent a reasonably even break through an organization which could battle for the common cause instead of wasting energy in internecine strife, the Independent Petroleum Association of America was formed. Wirt Franklin was its first president.

Such an association needed first of all a definite objective upon which its membership could unite. Nearly every member had a program or a plank for the association platform, but many of those planks were the very things which had kept the independents fighting among themselves for years while the majors largely controlled the industry. Wirt Franklin suggested a common purpose: "I believe American markets for crude oil should be kept for the American producer of crude oil."

That was something upon which all the independents could agree, and for the attainment of which they could work in unison. When the Humble Oil & Refining Company reduced the price of crude in Texas, and Carter Oil and Magnolia Petroleum followed suit in other states, Wirt Franklin called a meeting of his association, in Tulsa. The members cried loudly that the majors were again using the old weapons to restrict and destroy them. They maintained that the cut was unwarranted and had been made possible largely because of the fact that some of the richer companies were flooding the market with imported oil. The long battle for tariffs or excise taxes to restrict importation of oil was begun. It united the independents with a mutual interest that they had never known before. Wirt Franklin was its guiding spirit. Its ramifications and details are too complicated for even an attempted summary here, but it strengthened the organization of the independents throughout the country.

Before the NRA codes were proposed the oil industry seemed to be headed toward a point of comparative, if temporary, stability, although in the widespread economic depression it was not prosperous. It had within it, notably in Oklahoma, various men who had grown to wealth and practical power as independent oil operators. Each one of those was an asset to the association and to the industry in general. They had started operations years before with little more than keen intelligence, ambition and energy.

They had attained invaluable experience and large wealth. A leader among them was Frank Buttram. Other important figures of a different type were W. R. and W. E. Ramsey, brothers.

Buttram's story is along the Horatio Alger line which has become almost a theme of derision in a modern cynical generation. Perhaps nothing is more irritating to a young person trying vainly to gain a footing in business or profession today than to be told of the success of older men who started from scratch.

Despite such understandable cynicism there should be more comfort and inspiration in facts than in formulas. That young people and middle-aged people of today need and seek help and stimulation has been well demonstrated by the wide sales of so-called inspirational books in recent years. It is odd that a generation which so eagerly grasps conventional suggestions for a better life should scorn a factual record of the practical attainment of such a life. Odd, but true.

Yet the "poor boys who became famous" proved, before writers of inspirational books ever analyzed and precipitated the formulas of success, that opportunity awaits and will reward the application of intelligence.

Frank Buttram was one of those poor boys. Born in the Chickasaw Nation in 1886, the first of a large family of children, he grew through childhood close to the soil and close to the problems of the Last Frontier. His father and mother had come from Missouri in the earlier days. They were a poor family. Almost any family then attempting to wrest a living from the frontier was likely to be poor. The boy was accustomed to work. Fortunately for him in the years to come he was also accustomed to play. He led a normal balanced life for the time and environment in which he found himself. He had the qualities now urged upon us by all the writers of inspirational books—energy, ambition, intelligence, versatility.

The ambition made him wish to rise above the monotonous level of farm life. The intelligence suggested how he might do so. The energy supplied the needed force. The versatility brought him recreation to strengthen his efforts, and popularity among his associates. He worked his way through the Teachers' Institute at Tecumseh with high grades, and obtained a certificate to teach a country school.

When he applied for appointment to a district school not many miles from his home the trustee told him he could have it—for a seven-month term at thirty dollars a month. That was all right with him. It was an opportunity for the first step up from common labor. And when the trustee, a farmer, tried to sell him a buggy and harness on the ground that it would be more in keeping with his new-found dignity as schoolmaster than the old white horse he bestrode, he recognized another opportunity. The price was $35. His prompt action was another demonstration of intelligence.

"Pay me thirty-five a month instead of thirty, and I'll buy," he said.

The deal was made. He threw his saddle in the back of the buggy, harnessed the old horse between the shafts, and drove home, exultant.

Ambition, intelligence and energy maintained their pressure. Teaching and studying alternately, supplementing his cash with odd jobs whenever they could be found, but always finding time for athletics, debating and other college activities, he was graduated from the Normal School at Edmond. With enough university credits to give him junior class standing he entered the University of Oklahoma in 1909. Not so terribly long ago, 1909. This is quite a modern story.

In a year and one summer term Frank Buttram covered the two-year course required for his bachelor's degree. In the same period he gained such distinction as a college baseball pitcher that he struck out the famous Ty Cobb three times in one practice game which the Detroit Tigers were playing against Oklahoma U. New opportunity was opened by that accomplishment. But intelligence discriminates. Buttram turned down Cobb's suggestion that he could have a try-out with the Detroit team any time he wanted it. The advantages of a master's degree and what it stood for were more alluring.

Oklahomans were keenly alive to the possibilities of oil by that time. Dr. Charles N. Gould, director of the Oklahoma Geological Survey, impressed by Buttram's work as a student, appointed him chemist for the Survey. It was only a $50 a month job, but it relieved the drudgery of other jobs which had maintained him in college. It assured the winning of his master's degree, and at the same time familiarized him with important features of the state's geology. He made three studies and the Survey published three bulletins of importance by him. The third bulletin, on "The Cushing Oil Field," introduced Frank Buttram to the oil business. His analysis of the oils and sands of the Cushing field formed an important basis for further development of the entire mid-continent oil business. It also provided inspiration for Frank Buttram himself.

On the strength of information which he acquired in research for those bulletins, Buttram became interested in some different geological structures near the Cushing field. He interested S. W. Ohern, who had become state geologist, and A. P. Crockett, an attorney in Oklahoma City. The three managed to interest eastern capital, and organized the Fortuna Oil Company to test and exploit the projected new field.

It was not long before the Fortuna Company was able to sell four eighty-acre tracts for a million dollars in cash. The company promptly declared a $300,000 cash dividend. Frank Buttram received $30,000. He put $20,000 into a comfortable home in Oklahoma City and began to extend his interest in the oil business with the other $10,000.

From that time on he was never headed. As president of the Buttram Petroleum Company he extended his operations into the California and Texas fields. In the Texas field he found not only oil but opportunity to demonstrate his ability and determination in a manner which won him not only a fortune but the respect of oil men throughout the country.

He told me about it when I asked him for a story of difficulties encountered. "The public is more inspired by a story of difficulties overcome than by wealth attained without effort," I suggested. "Surely you must have found yourself up against a serious threat of defeat at times. What did you do?"

"Maybe my experience with the Texas company in the Corsicana field in Texas is a case in point," he said. "I had been operating in the Powell field. My first well there was good for 23,000 barrels a day. I knew that whole country pretty well, and I took a lease on what appeared to me to be a superior prospect at Corsicana. Soon after that I learned that the Texas company had leased all the ground surrounding mine and was planning to drill ten wells offsetting my holdings. That many wells would have stripped my ground.

"I did not want to drill at that moment. Prices of crude were not right. Transportation and marketing from the Powell field had been difficult. But the Texas company had the marketing facilities and the immense capital which probably would have enabled it to handle the Corsicana product and destroy the value of my holdings there. Could I, as an independent, afford to fight them? It was an important question for me. I gave it some intensive thought. If I gave battle and won it would be a tremendous help to my financial position and to my standing in the oil business. If I lost, it might be ruin. I figured my resources and decided to go ahead.

"I dispatched a telegram to the manager of the Texas company: 'If I do not get a wire from you tomorrow saying you will drill only one well, I shall drill ten. I can do it before you because I have ten rigs available and water available, and you have neither. If you do agree, I will give you two weeks start on the one well.'

"That was not a bluff. I would have gone through with it, though it would have been a desperate expedient which might have meant ruin if the field did not come in as anticipated. I waited very tensely you may be sure, for the reply. When it came I heaved a sigh of relief. The Texas agreed to my proposal.

"I gave them a three weeks' start with their well. Then I proceeded to drill. Fate intervened in my favor. Their haste spoiled their well. I beat them to production with a 10,000-barrel well, and took fifty times as much oil as the Texas company from an equal area."

That is a sample of one phase of the accomplishments of Frank Buttram. In 1926 he was voted Oklahoma City's "most useful citizen." He is a versatile man. He has served with distinction on the board of regents of his alma mater, endowed a loan fund for students who sought an education under financial handicaps similar to his own, proved himself a competent bank director with large interests in the Security National Bank of Oklahoma City and the Seaboard National Bank of Los Angeles, served as chairman of the Arkansas-Red River Flood Control Association and worked effectively upon the flood control problems of a dozen states. He has taken an active part in politics and philanthropies.

When the New Deal began the tremendous task of gathering all business and industry under the broad wings of its blue eagle, Frank Buttram was one of the men called as a competent representative of the oil business. The oil code, in the preparation of which he had a leading part through ten long months of constant labor, was a monument to his broad and accurate knowledge. He is still proud of that code, although I gathered that he is not proud of the New Deal.

Such men have made of Oklahoma City in little more than a decade one of the most substantial and impressive cities of the great Southwest. Its leaders and their characteristics are as varied as its resources. It began with the signal of a gunshot upon the Last Frontier less than fifty years ago. Through nearly forty years it grew in population and wealth from prairie boom town to shipping center, to state capital, to industrial importance. Still it was not impressive.

Then came oil. With greater speed and enthusiasm than it had known since the first exciting days of its original settlement it blossomed into the modernism of towering skyscrapers. Of these, one of the two which competed most spectacularly for the honor or advertising advantage of being "the tallest building" is the Ramsey tower. It is in truth a building to be proud of. It is more than that. It is a monument to the diversity of character and the breadth of opportunity which has brought the Last Frontier into the architectural lines of New York itself. The brothers who promoted it are as interesting in their accomplishments as any men who have taken a leading part in the building of the city.

W. R. Ramsey was out of town when I called to make their acquaintance. W. E. Ramsey was in his spacious office in the Ramsey Tower, just completing a telephone call to his wife at their winter home which he maintains on the Santa Monica bay sector of Los Angeles suburbs. The baby was well. The wife was well. Everything was fine. Mr. Ramsey hoped to get out there in a week or so. He was in excellent humor. He usually is, I had been informed by friends.

"A great salesman. He can take a map, a blueprint and a block of leases and sell enough royalties to drill an oil well, anywhere, anytime."

And why not? The Ramsey success is another story of "poor boys who became famous," as interesting as the stories already told, but quite different. It breaks away from the usual success story formula in that it dispenses entirely with the factor of education. When W. E. Ramsey happened to mention a period in his youth when he was temporarily employed as a kitchen mechanic to earn his board and a place to roll his blankets, I jumped to a conclusion based upon numerous success stories. "Working your way through college?" I asked.

He grinned, "My God, no! You flatter me. All the education I ever got you could pay for with a dollar and a half."

Apparently it was enough. According to an article in the *Daily Oklahoman* of February 13, 1927, shortly before the Oklahoma City field came in with a roar that stirred the world, the Ramsey brothers already owned oil leases on 235,000 acres of land in Oklahoma, Texas and Colorado in addition to richly productive holdings in the famous Signal Hill field at Long Beach, California.

W. E. Ramsey himself had found his first job as a shoe-store clerk in Ardmore, Oklahoma, at eighteen dollars a week. It seemed to offer slight promise for his potential abilities as a salesman. He knew the oil game only by reputation, but in 1913 the excitement of fortunes made in oil was familiar to everyone. The retail shoe business was distinctly limited. W. E. Ramsey decided to go into oil.

To attain success according to his theory of the foundation of success—"Know the game; work hard,"—he must first know the game. To do so he sacrificed half his income, and went to work as a laborer at $1.50 a day in the newly discovered Healdton field. After an apprenticeship at that he became a driller. And shortly he began to lease, and to sell interests in his leases for enough money to sink wells.

W. R. Ramsey in the same period had progressed from his first cash job as a ranch hand near Ardmore to a clerkship in a bank at Apache which paid him forty dollars a month. He had written insurance, tried out the real-estate business and established a personal credit which enabled him to borrow money to buy a small stock interest in the Guaranty State Bank when it opened in Oklahoma City. It was a short and natural step from banking to oil. He obtained three wildcat leases in Louisiana. W. E. Ramsey sold enough shares to finance the drilling of three wells. All three came in. The Ramseys were in the money.

But being in the money is not the only end, even though it may be the only aim, of the oil business. The Ramseys traveled over a veritable roller coaster of ups and downs after that. Their highest point probably is revealed in the Ramsey Tower. Their lowest, I do not know. Possibly it was in the failure of one of the greatest wildcat ventures in all the checkered history of oil-wildcatting. It was a venture in which they obtained control of 111,000 acres under lease in the Oklahoma Panhandle, and saw the entire project go up in smoke with their first test well.

Such is the story of oil upon the Last Frontier, and for that matter in almost every region in the world where petroleum has been discovered and produced. Luck and science, cautious investment and reckless speculation, honesty and fraud, errors and accuracy of judgment, cynical opposition and fanatical faith have played their part. Rewards and ruin have come alike upon the worthy and the unworthy, upon the just and the unjust.

Oil has built great cities—greater than gold has ever built, and apparently more permanent. Though the varied adventures of its development perhaps have been similar in every oil-producing state in this nation and throughout the world, Oklahoma I believe offers the most significant illustration of the astonishing effect of oil upon the civilization which produced it.

Witness the change from a territory virtually without government, overrun by bandits, restricted and retarded by a population of Indians of whom many were but slightly removed from savagery. Witness the change from the rambling ramshackle towns of Tulsa and Oklahoma City as they were only a quarter of a century ago into the great, clean, modern, enviable cities that they are to-day, combining most of the material advantages of a New York with the still adventurous and democratic spirit of the West, and offering a charm and hospitality suggestive of the days of the old South.

Oil did that; oil and its accompanying wealth used to its fullest possibilities by the men who discovered and produced it. The fact that Pennsylvania, California, Texas, Louisiana and other states have produced oil in vast quantities, in no way belittles Oklahoma's accomplishment. Not one has built upon oil as Oklahoma has built. Not one reveals the contrasts and drama of life before and after oil as Oklahoma reveals them.

There, in very truth, oil brought an end to America's Last Frontier.

(THE END)

BIBLIOGRAPHY

BIBLIOGRAPHY

Abbott, L. J., *History and Civics of Oklahoma.* Ginn & Co., New York and Boston, 1910.

Barnard, Evan G., *A Rider of the Cherokee Strip.* Houghton, Mifflin Co., New York, 1936.

Bass, Althea, *Cherokee Messenger.* University of Oklahoma Press, Norman, Oklahoma, 1936.

Boren, Lyle H., and Dale, *Who's Who in Oklahoma.* Co-Operative Publishing Co., Guthrie, Oklahoma, 1935.

Collings, Ellsworth, *The 101 Ranch.* University of Oklahoma Press, Norman, Oklahoma, 1937.

Conover, George W., *Sixty Years in Southwest Oklahoma.* N. T. Plummer, Printer, Anadarko, Oklahoma, 1927.

Foreman, Grant, *Indian Removal.* University of Oklahoma Press, Norman, Oklahoma, 1932.

Grinnell, George Bird, *The Fighting Cheyennes.* Charles Scribner's Sons, New York, 1915.

Hall, J. M., *The Beginning of Tulsa.* Privately published.

Harlow, Rex, *Oklahoma Leaders.* Harlow Publishing Co., Oklahoma City, 1928. *Oklahoma Yesterday and Tomorrow.* 1930.

Harman, S. W., *Hell On The Border.* Phoenix Publishing Co., Fort Smith, Arkansas.

Linderman, Frank B., *American.* The John Day Co., New York, 1930.

MacLeod, William Christie, *The American Indian Frontier.* Alfred A. Knopf, New York, 1915.

Raine, William MacLeod, *Famous Sheriffs and Western Outlaws.* Doubleday, Doran & Co., New York, 1929.

Rainey, George, *The Cherokee Strip.* Co-Operative Publishing Co., Guthrie, Oklahoma.

Sabin, Edwin L., *Wild Men of the Wild West.* Thomas Y. Crowell Co., New York, 1929.

Stansbery, Lon R., *Passing of the 3D Ranch.* Privately published.

NEWSPAPERS

Barde, F. S., Miscellaneous news articles. Collection, Oklahoma Historical Society, Oklahoma City.

328 THEN CAME OIL

Ahlum, Walter, "The Romance of Tulsa." Published serially by *The Tulsa World*, 1936.

The Daily Oklahoman, Oklahoma City. Special Editions, September 25, 1927; September 27, 1936. Numerous articles by Claude V. Barrow, Oil Editor. Oil development news items, etc.

Bartlesville Enterprise, Bartlesville, Oklahoma, "Statehood Edition," August 19, 1906. Articles: June 23, 1908, *et seq.*

The Enid Daily Eagle, Enid, Oklahoma, July 7, 1910.

The Evening News, Ada, Oklahoma. "Gov. Haskell Wants President to Explain His Bad Bargain For Osages," July 7, 1910.

The Muskogee Times, Muskogee, Oklahoma, May 5, 1903, *et seq.*

The Muskogee Democrat, Muskogee, Oklahoma, April 21, 1905, *et seq.*

The Muskogee Times-Democrat, Muskogee, Oklahoma, December 27, 1906, *et seq.*

Mueller, Harold L., "Tulsa's Titans." Published serially by *The Oklahoma City Times*, April 17, 1934, *et seq.*

News-State Tribune, Guthrie, Oklahoma, July 30, 1908.

The New York Times, New York, "Courts End Osage Indian Reign Of Terror," January 17, 1926; "Oklahoma Chooses Oil Before Beauty," February 16, 1936.

The Kansas City Star, Kansas City, Missouri, "Oklahoma's 'Coal Oil Johnny,' " December 21, 1913.

Oklahoma City Times, August 19, 1908; December 5, 1928.

Oklahoma State Capital, Guthrie, Oklahoma, August 8, 1908, *et seq.*

The Sunday News, New York, "Osage Terror," December 19, 1926.

The Seminole Producer, Seminole, Oklahoma, James T. Jackson, Publisher. Tenth Anniversary Edition, July 16, 1936. Special Edition, February 28, 1937.

The Tulsa Daily World, Tulsa, Oklahoma. Numerous articles by Paul S. Hedrick, and by others, on oil news and personalities.

MAGAZINES AND PAMPHLETS

American Magazine, "Overnight Millions," by Cora Miley.

Clinton, Dr. Fred S., *Journal of The Oklahoma State Medical Association.*

My Oklahoma, Oklahoma City, Parker La Moore, Editor. April, 1927, *et seq.*

The Oil and Gas Journal, Tulsa, Oklahoma. Diamond Jubilee Edition, August 23, 1934. Articles by Thomas F. Smiley, 1929.

The Oil Trade Journal, "Them Were The Days," by Col. H. L. Woods, October, 1925; "Some Legends of Petroleum," by Harry Botsford, October, 1926; Article by Charles W. Grimes.

The Oil Weekly. Gulf Publishing Co., Dallas, Texas, April 23, 1926.

Oklahoman Almanac, Oklahoma City, 1908-9-16-30.

Oklahoma. Official Publication of Oklahoma City Chamber of Commerce. Petroleum Number, November 26, 1936.

Outlook and Independent, "Oil Hells of Oklahoma," by Earl Sparling, February 11, 1931.

Review of Reviews, "Black Gold in Oklahoma," by Howard Florance, September, 1929; "Skelly's Ten Year Progress," by John Temple Graves II, October, 1929.

Stanolind Record, Chicago, Illinois, October, 1932.

The St. Louis Republic, St. Louis, Missouri, August 11, 1912.

Tulsa Chamber of Commerce bulletin on chronology of oil development in Oklahoma from 1901.

INDEX

Ackerley, tool of Payne, 45
Adams, Lewis, gave lease to Wick, 132
Adams, Thomas J., gave lease to Wick, 132
Adams, Wash, gave lease to Wick, 132
Adobe Walls, Indians killed hunters at, 55
Alabama, 20
and the Indians, 12
Alder Gulch, Montana
compared with Oklahoma as frontier, 89
compared with Seminole, 253
American, 34
American National Bank, Tulsa, taken over by Exchange National, 236-238
American Revolution, 12
Anadarko, Oklahoma, terminus of Chisholm Trail, 41
Apple, S. A., and Wirt Franklin, 312
Apple-Franklin Oil Company, and Wirt Franklin, 313
Arapahoe Indians, 78
established on reservation, 28
Archbold, John D., and discrimination against independents, 208
Arkansas, 19
Western Cherokees in, 13
Arkansas City, 71
and opening of Cherokee Strip, 80
Payne set out from, 45
Arkansas River, 12
Keeler got drilling outfit across, 116-117
Atkins, Minnie, and Tommy Atkins lease, 203*ff*

Atlantic & Pacific Railroad, *see* St. Louis & San Francisco
Aylesworth, Secretary of Dawes Commission, and Sue A. Bland well, 135

Bailey, Clarence, filed suits against various oil companies, 272
banking, and oil, *see* Chapter XVI
Barde, Fred S., gave news of Sue A. Bland well, 136
Barnard, E. G., *Rider of the Cherokee Strip, A*, 83, 88
cited on life on frontier, 88*ff*
Barnard ranch, famous, 125
Barnett, Jackson, wealthy "incompetent" Indian, 156, 229-230
Barnsdall, T. N., 181
and renewal of Osage blanket lease, 151*ff*
connected with I. T. I. O. and Standard, 150-151
Barnsdall Corporation, Waite Phillips' holdings turned over to, 213
Barrel House Sue, Seminole bad woman, 255
Barrow, Claude V.
cited on Moran, 284
cited on Slick, 200
Bartles, Jacob H.
owned mill on Caney River, 118
sold right of way to Santa Fe, 119
Bartlesville
banks of, 210, 211
celebration upon Oklahoma's becoming a State, 160-161
first productive well in, 198
oil metropolis, 117, 118-119

Starr, Henry, and "Cherokee Bill," 98
Starr, Sam, husband of Belle, 64-66
Starr, Tom, Belle and husband made home with, 63
State Capital, Oklahoma, quoted on Haskell vs. West, 207
Steelsmith, Amos, 116, 231
Stillwater, and opening of Cherokee Strip, 80
Stink, John, *see* Ho-tok-moie
Strother, O. D., and the Seminole field, 246*ff*
Sudik, Mary
  characteristics of, 296*ff*
  well named for, 296
Sudik, Vincent, oil on land of, 293*ff*
Sue A. Bland well, 198

Teapot Dome oil scandal, 197
Tennessee, and the Indians, 12
Tennessee River, Cherokees encountered white traders on, 13
Texas Company, 142, 232, 284
  and Buttram, 318
Texas Panhandle, 19
Texas Pipe Line Company completed outlets to gulf, 167
Thomas, United States Deputy Marshal, went after the Dalton gang, 123
"3-D" ranch, described by Stansbery, 125*ff*
Tidewater Oil Company, 232
Times-Democrat, Muskogee
  cited on shrewdness of Charley the Creek, 163
  quoted on high-pressure lease buyer, 163-164
  quoted on righteous indignation of territory over removal of Osage blanket lease, 154-155
Times, New York, reported estimate of Cosden's wealth, 176, 177-178
Times, Oklahoma City
  cited on Exchange National of Tulsa, 233

Times, Oklahoma City—*Continued*
  cited on Sinclair, 198, 199
  quoted on Cosden, 178
  quoted on Haskell's tie-up with Standard, 207
Tommy Atkins lease, 183
  difficulties over, 203
Tonkawa, Oklahoma oil field, 245, 246, 281
Tonkawa Indians, part of reservation opened, 78
Tonopah, Nevada, and Jim Butler, 203
"Trail of Tears, The," exile of the Cherokees, 12
Tulsa, Oklahoma
  and Cosden, 175*ff*
  banks of, 210
  beginnings of, 129
  description of, 166-167
  hotels in, 166, 167
  in oil boom caused by Sue A. Bland, 136
  oil capital, 165-167, 234
Tulsa Chamber of Commerce, Skelly director of, 194
Twain, Mark, *Adventures of Tom Sawyer,* 37
Twin State Oil Company, 261

Ufer, F. B., organized Exchange National, 215
Unassigned Lands
  boomers in, 44
  bought from Creeks and Seminoles, 44, 71
  opening of, 126
  Payne in, 44*ff*
  rush to, 71*ff*
Union Pacific Railroad, 39

Van Brunt, Mrs. E. W., 106
Van Buren, Arkansas
  Childers broke jail at, 53
Vaughn, Charles, murdered, 268
Vennoy, United States deputy marshal, and John Childers, 52